The Voluntary Principle in Conservation

The Farming and Wildlife Advisory Group

The Voluntary Principle in Conservation

The Farming and Wildlife Advisory Group

Graham Cox
Philip Lowe
Michael Winter

PACKARD PUBLISHING LIMITED
CHICHESTER

Copyright © Graham Cox, Philip Lowe & Michael Winter

First published in 1990 by Packard Publishing Limited,
16 Lynch Down, Funtington, Chichester, West Sussex PO18 9LR.

All rights reserved. No part of this publication may be reproduced, stored in a retrieval system or transmitted in any form or by any means, electronic, mechanical, photocopying, recording or otherwise, without the prior permission of the publisher.

ISBN 1 85341 041 1 (hardback)

Phototypeset in Palatino by Barbara James, Rowlands Castle, Hampshire PO9 6EJ.

Printed and bound in the United Kingdom.

Contents

CHAPTER 1	*Introduction*	1

CHAPTER 2	*The Silsoe Exercise and the Origins of FWAG*	
	Introduction	8
	Lindsey Project for the Improvement of the Environment	10
	Background to Silsoe	12
	Farming and Wildlife: a Study in Compromise	14
	After Silsoe	17

CHAPTER 3	*The Development and Internal Politics of National FWAG*	23
	Introduction	24
	Farming and Wildlife Exercises	25
	The Struggle for Survival: Funding for FWAG	32
	FWAG in the Political Limelight	35
	Promotion of a National Advisory Service for Farm Conservation: FWAG and the Countryside Commission	49

CHAPTER 4	*The Establishment and Development of County Groups*	63
	Introduction	64
	MAFF Initiatives	64
	How to Reach Local Farmers?	67
	Promoting Local Groups	71
	Developing the Advisory Capacity of Local Groups	75
	Expanding the FWAG Network	81

CHAPTER 5	*The Workings of County FWAGs*	85
	Introduction	86
	Composition and Organization of County FWAGs	86
	Key Honorary Officers	91
	Advisers	93
	Advice to Farmers	97
	Effectiveness of Advice	104
	Conclusions	105
CHAPTER 6	*Stony Ground: FFWAG in Wales*	109
	Introduction	110
	The FUW and Conservation	111
	The Dinas Conference: a False Start	113
	From Dinas to FFWAG	116
	The Wales Countryside Forum	118
	Montgomery FFWAG	119
CHAPTER 7	*Fertile Prospects: FWAG in Wiltshire*	127
	Introduction	128
	Establishing a County Group	129
	A Chalklands Re-run?	133
	A Chairman and a Philosophy	136
	Trees and Grants	142
	Putting a 'Bod in the Field'	144
	The Shape of a New Era	149
	Patterns of Accommodation	154
	Consolidation and Advance	164
CHAPTER 8	*An Uncertain Future*	173
	The Voluntary Principle	174
	FWAG and the Countryside Commission: the Tension between Client and Sponsor	178
	FWAG and ADAS: Partnership or Rivalry?	182
	FWAG Prospects	186

Appendix:	FWAG information leaflets	190
References		191
Index		195

Figures

4.1	Growth of County FWAGs and Farm Conservation Advisers	82
5.1	Organizational Representation on County FWAGs	87
5.2	Countryside Advisers Supported by Countryside Commission, 1986	99

Tables

5.1	Most Common Requests for Advice to County FWAGs with an Advisory Officer	100
5.2	Most Common Requests for Advice to County FWAGs without an Advisory Officer	102
5.3	Non-Business Meetings and Events Organized by County FWAGs during 1985	104
5.4	Numbers and Percentages of Farmers Implementing Advice by Region	105
7.1	Sources by which Contact between Wiltshire FWAG and Farmers was made	165
7.2	Subject of Requests for Advice from Wiltshire FWAG, 1985-1988	166
7.3	Responses for Advice from Wiltshire FWAG, 1985-1988	167
7.4	Management Type and Hectarage of Farms Visited by Wiltshire FWAG, 1985-1988	169
7.5	Enterprise Type and Farm Size by Hectares of Advisory Visits Made by Wiltshire FWAG	170

Illustrations

Grateful thanks are expressed to the following people and organizations for the use of their photographs: Ruth Tittensor for the cover and lead-in pages to Chapters 1, 2 and 6; the Countryside Commission for the title pages and Chapters 4 and 5 (the last courtesy of the Gloucestershire County Planning Department); the Farming and Wildlife Trust for Chapters 3 and 7; and Charles Watkins for Chapter 8.

Acknowledgements

Debts quickly accumulate during the preparation of studies such as the one presented here. We are grateful to The Economic and Social Research Council whose grant D00232072 made the research on which the book is based possible. Our work had the support of the National Farmers' Union, the Country Landowners' Association and a number of other government and voluntary agencies concerned with agriculture and conservation. In addition to the valuable assistance we received from many active FWAG members we wish particularly to acknowledge the permission to conduct archival work which was readily granted both by Patrick Leonard and Keith Turner of the Countryside Commission and by Eric Carter, National Adviser to FWAG.

FWAG members in the counties of Essex, Montgomery and Wiltshire gave generously of their valuable time and enabled us to conduct extensive interviews and, in addition, the secretaries of most of the county FWAGs kindly completed and returned our postal questionnaire. We also received considerable help from officials at the Countryside Commission, the Nature Conservancy Council, the National Farmers' Union, the Ministry of Agriculture, Fisheries and Food, the Farmers' Union of Wales and the Welsh Office Agriculture Department.

During the later stages of our work a number of individuals kindly read and commented upon draft chapters and we are especially grateful to Sir Derek Barber, Eric Carter, Wilf Dawson, Jim Hall, David Lea, Professor Norman Moore, Bill Ratcliffe and Colin Small for the insights they were able to offer. We do not, of course, seek in any way to implicate them in what we have written and we can only repay their courtesy by emphasizing that the responsibility for any errors which may remain rests with us alone.

Graham Cox, Philip Lowe, Michael Winter

CHAPTER 1
Introduction

With the 1981 *Wildlife and Countryside Act*, nature conservation in Britain may be said to have come of age. Certainly the Act had many widely acknowledged weaknesses, some of which were remedied in amending legislation in 1985, but the debate and interest which it engendered both during and after its stormy passage through Parliament did much to bring conservation on to the centre stage of British politics. In the immediate aftermath much attention was devoted to the majority of the farmed countryside, not specifically covered by the Act. Agricultural leaders and ministers were quick to emphasize that the principles of goodwill and voluntary co-operation embodied in the Act were equally applicable to the wider countryside. In particular the rapid emergence of county Farming and Wildlife Advisory Groups in the late 1970s and early 1980s was perceived as evidence of a concern for conservation in the farming community. Since then, these FWAGs have been presented as the best available vehicle for demonstrating the capacity for farmers and conservationists to work together in harmony and as a means by which farmers themselves might be encouraged to adopt conservation practices in their farming.

National FWAG was formed in 1969 by the National Farmers' Union (NFU), the Country Landowners' Association (CLA), the Ministry of Agriculture, Fisheries and Food (MAFF), the Royal Society for the Protection of Birds (RSPB), the Society for the Promotion of Nature Reserves (SPNR, later to become RSNC), the British Trust for Ornithology (BTO) and the Nature Conservancy (NC). The aim was to bring together agriculturalists and conservationists to promote mutual understanding and co-operation. Since the mid-1970s a complete network of county FWAGs has been set up to involve local farmers and landowners. Their avowed principle is that conservation and modern farming need not be incompatible and that the loss of wildlife habitats through agricultural intensification can best be ameliorated by encouraging farmers to modify their practices through the provision of appropriate advice and encouragement.

FWAG is the prime expression of the voluntary principle in conservation. This has two components. One is, positively, to stimulate and broadcast amongst farmers and landowners a social ethic concerning stewardship of the countryside, including the protection and enhancement of natural diversity and beauty within the context of modern farming practice and estate management. The other component is an ideological defence of the autonomy of farmers and landowners from statutory controls, through an emphasis on the paramount need to retain

Introduction

their goodwill and voluntary co-operation if workable remedies to conservation problems are to be found.

Given the centrality of the voluntary principle to current conservation policy and the key role of FWAG as an agent of the voluntary principle, this book presents a study of that organization, analysing how it has set about formulating and disseminating a pragmatic conservation ethic and how, through it, a political strategy has been pursued to defend and justify a voluntary approach to conservation. The study was conceived as an investigation into the development of a particular public-private policy initiative, its associated ideology and its potential implications for promoting the reconciliation of agricultural and conservation objectives.

The research for the book was undertaken between 1984 and 1988 and was supported by a grant from the Economic and Social Research Council. Our first objective was to put together an account of the emergence of FWAG and of the importance in its development of the various agencies and interest groups which are members of the organization. Accordingly we undertook a detailed study of published and unpublished material – including in-house journals, internal memos, minutes, correspondence – subject to the degree of access granted by the different agencies and groups in membership. The material was checked and supplemented by interviews and correspondence with some of the people who have been most closely and actively involved. The depth and richness of this material has allowed us to present a well-documented history of FWAG. In Chapter 2 we trace its genesis to initiatives which arose out of the 'Countryside in 1970' conferences in the 1960s and from a number of meetings between key personnel in MAFF and the RSPB. Crucial to this development was the Silsoe Exercise, an attempt to show farmers through practical demonstration how the needs of conservation and modern farming might be reconciled. Not only did the Silsoe organizing committee provide the nucleus for the subsequently formed national FWAG committee but it also provided the model for a number of other events throughout the country and several of these in turn spawned local FWAG groups.

FWAG's development, covered in Chapter 3, was, nonetheless, slow and uncertain. Not until the period after the *Wildlife and Countryside Act* did it begin to achieve a nationally recognized profile and develop a wide coverage of local groups (the evolution of the network of county FWAGs is the subject of Chapter 4). Progress during its formative years was hampered in particular by lack of finance, but also by a degree of inter-agency rivalry. The rivalry came to a head in the mid-1970s when its future seemed particularly precarious. Its full-time national advisor, Jim Hall, who largely personified FWAG throughout the 1970s, was at one stage given a renewal of contract for just six weeks, so desperate was the financial crisis. Into that climate came the Nature Conservancy Council (NCC) and, more importantly, the Countryside Commission. Both were prepared to save FWAG only if certain conditions relating to its organization and objectives were met. Of paramount importance to the Commission was its long-standing and, at times, acrimonious dispute with FWAG concerning the degree to which FWAG should broaden its wildlife remit to include landscape considerations and the professionalism of the county groups in giving conservation advice.

The charting of these disputes has enabled us to learn a great deal not only about the specific development of FWAG but also the nature of inter-agency relationships within a policy community or, more accurately, where two separate policy communities intersect – in this case conservation and agriculture. In certain respects, FWAG is a typical product of long-established corporatist arrangements elsewhere in the agricultural industry. These have been based on, and sustained by, the industry's desire to remain free of controls except where these can be administered by the industry itself or by the Ministry of Agriculture. Increasingly the farming and landowning lobbies used FWAG to demonstrate a commitment to conservation and have sought thereby to further their strategies of resisting the imposition of external controls on the industry.

The interests of the farming and landowning lobbies and the official conservation agencies have coincided in building up the county FWAGs as vehicles for providing advice to farmers. To understand how these groups operate and to assess their potential for fulfilling this role we distributed a postal questionnaire to all county FWAG secretaries in England and Wales seeking information on organisation, activities, finance, advisory work, membership, and so forth (see Chapter 5). Each county FWAG is made up from one to several dozen individual members, a significant proportion of whom are practising farmers with the rest being drawn from the various conservation and agricultural bodies in the county: ADAS, the NCC, Countryside Commission, RSPB, county trust for nature conservation, Council for the Protection of Rural England (CPRE), county planning department, and so forth.

In order more fully to appreciate the ideology of compromise FWAG promotes we felt it necessary to discern the views, not only of senior FWAG spokesmen, but also of FWAG members in the counties. We accordingly selected two counties in which to undertake in-depth interviews with FWAG committee members. The counties were chosen for the contrasts they presented both in terms of farming patterns and the success of the local FWAG. In the case of Montgomery we had an upland county with very small farms and a relatively weakly developed county group, and it served to illustrate some of the problems encountered in establishing Farming, Forestry and Wildlife Advisory Groups (FFWAGs) in Wales (see Chapter 6). Wiltshire, on the other hand, is noted for its productive arable and mixed agriculture, dominated by larger farms. Its FWAG is widely recognized as one of the most successful in the country with an active committee and adviser (see Chapter 7). In spite of obvious differences between the two committees we found a remarkable degree of concurrence, especially among farmer members, of views on both FWAG and wider countryside issues. The analysis of the semi-structured interviews, which lasted anything from one to three hours, provides a valuable insight into how FWAG members on the ground perceive their roles.

The final phase of the project was to consider the policy implications of our work and the future direction of FWAG. This has involved us in the dual role of making informed speculation and suggestions for reform. Two things are particularly important to FWAG's future. The first is finance: currently FWAGs' main cost is employing advisers in the counties. The Countryside Commission agreed to finance 40 per cent of the cost of these posts for the first three years, the Farming & Wildlife Trust (a national charity set up by FWAG expressly for this purpose) a further 25 per cent and local sources the remainder. The crucial question is whether the Trust and local sources will be able to take over from the Countryside Commission. Second there is the question of relations with the official Agricultural Development and Advisory Service (ADAS). Previously, ADAS provided unparalleled local support 'in kind' for FWAGs, providing a secretariat, helping to organize events and so on. However, in recent years such support has come under strain. The servicing needs of FWAGs have grown as advisers have been appointed and ADAS itself has been put under considerable pressure through cuts in its budget. Furthermore as a means of diversifying its role in the face of political hostility MAFF is seeking to promote ADAS itself as the primary source of conservation advice for farmers. There are clear conflicts of interest and loyalty here, and the final chapter examines these in considering what the prospects are for ensuring effective farm conservation advice.

CHAPTER 2
The Silsoe Exercise and the Origins of FWAG

Introduction

Tracing the origins of any kind of political or institutional arrangement is beset with difficulty. Inevitably the benefit of hindsight can bestow upon certain events a significance which was, perhaps, barely apparent to the participants at the time. Moreover, a process of generating a preferred account of the past invariably accompanies the growth of any organization. So, whatever the contemporary import of events, and however apparent the progress towards a certain goal, it is even more dangerous to assume that the past intentions of individual participants were uniform and necessarily directed towards the eventual outcome. Such caveats notwithstanding, it is evident that the formation of FWAG was not an isolated, unheralded initiative. Rather it was a direct outcome of processes set in motion by the Silsoe Exercise, which in its turn has precedents in some of the initiatives of the 'Countryside in 1970' Conferences held at intervals through the 1960s (the main conferences were in 1963, 1965 and 1970).

The formation of FWAG was also associated with the changing general climate of opinion on environmental matters. The burgeoning of environmental concern, from the mid-60s onwards, encompassed the gloomy predictions and radical remedies emanating from the nascent ecology movement and the conservative preservationism of middle-class amenity societies (Brooks *et al.*, 1976; Lowe and Goyder, 1983; Sandbach, 1980). In some quarters notions such as 'science', 'progress' and 'development' came under increasingly critical scrutiny as part of a rejection of the values of 'industrial society'. Other sectors of the environmental movement retained a commitment to scientific and technological rationality. The notion of 'conservation' as 'rational resource management' was promoted specifically by the Nature Conservancy, but was promulgated also by other agencies with statutory responsibilities for natural resources, including the Forestry Commission, the newly created Countryside Commission and many county planning authorities in their new-style structure plans. The concept implied that environmental problems were susceptible to informed planning, the mutual accommodation of affected interests and the search for technical solutions – tasks which presumed a key role in the regulation of resource use for planners and scientists. The often fractured dialogue and tension between these contrasting approaches to environmental problems was and, to some extent, remains a major component of environmental politics.

Nevertheless, the dominant perspective of the 'Countryside in 1970' conferences was conservation, with consensual and technocratic overtones, though rural preservation interests were also well represented. Inevitably, agricultural change was a focus of attention. From a conservation perspective, the intensification and expansion of agriculture strained the capacity of the countryside to sustain a variety of uses and generated conflicts with other rural interests. From a preservationist perspective, there was concern at the impact of modern farming techniques on the traditional landscape and wildlife of the countryside. The key ecological concerns expressed were the impact of pesticides on food chains and problems of soil deterioration. Whereas the 'Countryside in 1970' conferences, with their royal patronage and inclusion of a range of interest groups and statutory authorities, were intended to stimulate public interest and political action, FWAG was aimed at a narrower and more specialist audience. Like the wider conferences, however, it embodied faith in progress through rational discussion, and a search for compromise through a combination of appropriate management and education.

Rather than deploring countryside changes, many at this time saw them as the consequences of desirable progress presenting opportunities to be grasped, a mood particularly well expressed in the book by Nan Fairbrother, aptly titled *New Lives, New Landscapes,* published in 1970. Fairbrother sought a "new landscape framework". Her discussion of agriculture praised its efficiency and embraced with enthusiasm the "openness and space" of the new farming landscapes. "Our new-style farmland is a living and thriving landscape efficiently adapted to the modern world and productive as it never has been before, and though we may regret the old countryside which it is ploughing away we must still admire its vitality" (p.227). In retrospect, her outlook seems all too naïvely optimistic about the capacity of the countryside to adapt to major technological change while still providing an interesting and attractive environment. Its emphasis on landscape to the virtual exclusion of wildlife considerations was, moreover, a crucial failing. Her approach was, however, consistent with the prevailing optimism and belief in the merits of responding creatively to the challenge of technological change. FWAG itself was a product of such dominant presuppositions.

A number of points about the 'Countryside in 1970' conferences need to be emphasized. Principally, the conferences raised, for the first time in such depth, many of the issues facing the contemporary countryside. Second, they provided a hitherto unprecedented opportunity for an exchange of views between a wide range of rural interest groups and statutory authorities. Third, by the late 1960s, after several years of conferences and discussion groups since the first meetings in 1963, there was a feeling that talk must give way to action. It was further felt that the Conferences' message must be taken to a wider audience, both in terms of educating the public as to what was at stake and in increasing awareness on the part of those whose professional activities affected the future of the countryside. There was concern that the whole thing was becoming a talking shop and although the representatives and officials of various interests and agencies were involved, the practical people, the field staff, were not. One of the results of these concerns was the passage of the 1968 *Countryside Act*, notable for the establishment of the Countryside Commission. The Lindsey Project for the Improvement of the Environment (LPIE), meanwhile, was the first step taken to seen if the Conferences' spirit of co-operation could work on the ground.

Lindsey Project for the Improvement of the Environment.

The LPIE was sponsored by the 'Countryside in 1970' Standing Committee, the Carnegie United Kingdom Trust, Lindsey and Holland Rural Community Council and Lindsey County Council. Additional members on the steering committee were drawn from the Lincolnshire Trust for Nature Conservation and the Countryside Commission. Though unconnected, in some respects it was a forerunner of the Silsoe Exercise, the immediate precursor to FWAG, and certainly many of those involved in the formation of FWAG had some involvement in, or were aware of, the LPIE.

Its location in Lincolnshire was significant. The county was the home of one of the earliest established (1948) of the county trusts for nature conservation. Indeed the Secretary of the Lincolnshire Trust in the 1950s, A.E. Smith, had been largely instrumental in establishing the national network of county trusts (Lowe and Goyder, 1983), and in moving the headquarters of the Society for the Promotion of Nature Reserves (SPNR, now the Royal Society for Nature Conservation) to Lincolnshire to act as the national umbrella body for the county trusts. As one of the epicentres of the post-war revolution in arable farming techniques, Lincolnshire was at the forefront of the debate on the relationship between modern farming and conservation. Of particular importance was the contribution of Dick Cornwallis, a former Chairman of the Lincolnshire NFU and an active member of both the Lincolnshire Trust and the Executive Committee of the SPNR. Cornwallis was far more than a farmer with a political grasp of the need to accommodate conservation interests in agricultural developments. He was a keen ornithologist and campaigner for conservation, and perceived that many of the contemporary developments in farming practices and techniques were inimical to wildlife.

In 1969 Cornwallis published a paper in *Biological Conservation*, in which he discussed agricultural change, particularly regional specialization, and its effect on wildlife. Much of the change he regarded as inevitable but he maintained that, within a highly productive and technically efficient agriculture, room could be found "to maintain a richness of variety by conscious development of the odd, uncultivatable corners that exist on every farm". From this planned development Cornwallis confidently expected "a new landscape" to emerge. His paper was both timely and influential, representing an important contribution to the growing body of thought which stressed the compatibility, with sufficient foresight, of agriculture and conservation. As well as influencing the strategy of LPIE, he also took part in the meetings which preceded Silsoe. But for his untimely death in 1969 he would undoubtedly have become a leading figure in the emergence of FWAG. The LPIE took place between September 1967 and August 1970. Its main objective was defined as follows:

> To demonstrate the way in which voluntary bodies, individually and collectively, could collaborate both between themselves and with the statutory bodies to achieve the preservation and promotion of a high quality environment.

The aim was to achieve closer collaboration between the various organizations concerned with the rural environment, which was one of the strongest messages coming out of the 'Countryside in 1970' conferences. The Project Officer, John Leefe, was seconded by the Forestry Commission from his post of District Forest Officer for South Lincolnshire and Rutland. Previously, for ten years, he had been lecturing in forestry. This naturally led to an emphasis in the Project on promoting tree planting on farms: something that was carried on very strongly in the work of FWAG. Also carried over was an emphasis on simple, cheap, and above all, practical advice to farmers. Leefe produced advisory leaflets which were issued to farmers by the Lindsey National Agricultural Advisory Service (NAAS) and Agricultural Executive Committee (Leefe, 1968). FWAG's first leaflet *Farming with Wildlife* was similarly designed, and, indeed, reproduced some of Leefe's diagrams of tree plantings in field corners (Leefe, 1970, p.23). One of the LPIE achievements was a 'Trees on the Farm' seminar organized in conjunction with Eric Carter, then NAAS County Agricultural Adviser for Lindsey, including representatives of the MAFF Agricultural Land Service and the NAAS, the Forestry Commission, Lindsey County Council, the NFU, the CLA, the Lincolnshire CPRE Trees Committee, the Kesteven Tree Society and the Lincolnshire Trust. The main aim of the group was to give advice and assistance on tree planting to farmers.

There are a number of respects in which the LPIE might be said to have prefigured FWAG:

(1) Its commitment to the co-ordination of statutory and voluntary bodies;
(2) Its emphasis on active, practical conservation, 'the improvement' of the environment through tree planting and other schemes, rather than the preservation of existing features;
(3) The emphasis on the provision of advice and education;
(4) Its acceptance of the inevitability of changes to the farmed landscape; on the removal of hedgerows, for example, Leefe urged "we have to accept this trend" (Leefe, 1970, p.13);
(5) Above all the LPIE was premised on a voluntary approach which relied on, and sought to engender, goodwill; having described the importance of the agricultural industry in the county, the LPIE report concluded that "the support of the farming community for measures leading to the conservation of landscape features and wildlife and the development of recreational facilities must be obtained" (Leefe, 1970, p.7).

The conclusions of the project, however, concentrated largely on the role of local authorities and voluntary bodies in conservation, with many suggestions put forward on how they might work together more effectively. The possibility that the farming community and the Ministry of Agriculture might also become active partners in this endeavour was not addressed. This was to be the particular contribution of the Silsoe Exercise.

Background to Silsoe

The Silsoe Exercise was the idea of Eric Carter. It arose out of a series of meetings in 1967 and 1968 between leading agriculturalists and conservationists brought together by Derek Barber, the NAAS County Agricultural Officer for Gloucestershire and David Lea, the Deputy Director and Reserves Manager of the RSPB. The meetings were initiated by Lea who, as a former farmer, was anxious to bring together agricultural and conservation interests to discuss the implications of the sweeping changes occurring within farming. The first was held at the RSPB's headquarters at the end of June 1967. Besides Barber, Carter and Lea, it included Peter Conder, Director of the RSPB, Dr Norman Moore and Dr Bruce Forman of the Nature Conservancy, Kenneth Williamson of the British Trust for Ornithology, Dick Cornwallis from Lincolnshire, and John Trist, the County Agricultural Officer for Suffolk. Discussion roamed widely over what was in store for the countryside ('Take a Pride in Your Landscape', Minutes of the Meeting, August 1967, FWAG Archive).

The participants showed a deep, perhaps in retrospect, almost an exaggerated, awareness of the changing nature of agriculture. Not only were fields becoming larger but production was being transformed. The meeting envisaged the in-door production of all lowland livestock with grass-drying and large-scale vegetable production as the methods of feeding. Conversely, hill-farming would decline with land tumbling out of production. The tone reflected the scientific and futuristic optimism of policy makers in the late 1960s. The importance of naturalists having an understanding of the economic and technological forces behind agricultural change was stressed. Particular emphasis was accorded to hedgerow removal. The economic necessity for much removal was accepted and fields of 30 to 40 acres were seen as the optimum size. An emphasis was placed, therefore, on the need to create alternative features, for instance, by planting trees in field corners. Another issue which exercised the group was the use and effects of pesticides.

The potential of the NAAS was identified in promoting wider attention among farmers to the needs of conservation and giving encouragement to sympathetic farmers. It was recognized that, for many, wildlife alone might not have sufficient appeal, but that farmers might be persuaded to adopt conservation practices if the advantages for game and possible improvements to the 'visual aspect' of farms were highlighted. The meeting concluded that pilot conferences should be run in the counties, organized by NAAS, and that a weekend demonstration survey of a farm be held with a view to formulating a plan that would incorporate conservation and agricultural interests. David Lea has commented:

> As our discussions developed we took the strong view that we had to get some of the people who formed opinion in the farming world

over to our side so that their conservation activities could be written about in the farming press and that they would become people to copy. (Personal Communication, 17 November, 1986)

The NFU and the CLA were approached and asked to nominate members to serve on an organizing committee for a national farming and conservation event. The RSPB staff took particular interest in promoting the project. When the *Farmers' Weekly* offered to sponsor the proposed weekend conference, David Lea recommended to the RSPB Council that the offer should be declined and that the Society itself should act as sponsor. In the event an outbreak of foot-and-mouth disease delayed the implementation of these plans by more than a year.

Farming and Wildlife: a Study in Compromise

The Silsoe Exercise, in July 1969, brought together one hundred farmers and conservationists from different parts of the country, at the Agricultural College at Silsoe in Bedfordshire. It, too, was sponsored by the Standing Committee of 'The Countryside in 1970' conferences. The aim of the exercise was to find practical ways of reconciling the interests of farming and wildlife through management strategies designed to facilitate profitable modern farming methods compatible with the needs of wildlife (Barber, 1970). Whatever the technical merits of its findings – and some ecologists have certainly been critical of the principles involved (e.g. Green, 1971; 1975) – the event represents something of a milestone in the dialogue between agriculture and conservation. It spawned a series of similar exercises in different parts of the country and, above all, led directly to the formation of the Farming and Wildlife Advisory Group.

It is important to understand the nature of the Silsoe Exercise and the reasons why it attracted both a high degree of favourable publicity and acted as so effective a stimulus for subsequent developments. The essence of the exercise was that "the leaders of the farming and conservation establishments" (Barber, 1986) across the country met together over two days not only to discuss, but also to look in detail at a particular area and to prepare possible plans for its future development. Some 160 ha in Hertfordshire were chosen for study, comprising two adjoining farms, one run by *Farmers' Weekly*, the other by a friend of Barber's, Clifford Selly. The land, which retained a large number of hedgerows and traditional features, had been the subject of detailed bird censuses by the BTO over several years. In addition, the plants, fungi, lichens, insects and mammals had been surveyed for the event by the Nature Conservancy and local naturalists. Speakers and chairmen at the conference included James Fisher, Deputy Chairman of the Countryside Commission, Emrys Jones, then Chief Agricultural Adviser to the Minister of Agriculture, Walter Smith, Director of NAAS, and Martin Holdgate, Deputy Director of the Nature Conservancy. The President of the National Farmers' Union, Henry Plumb, was prevented by illness from chairing one of the sessions. Other speakers included David Lea, Norman Moore, Derek Barber and Eric Carter.

The opening address was given by Sir John Winnifrith, Director of the National Trust and former Permanent Secretary at the MAFF. He urged the necessity for conservationists and farmers to work together and overcome their differences which, he argued, were minor compared with their mutual opposition to urban growth:

> The common enemy is the urban industrial interest. It surely is a fact that with the vast predominance of the urban interest, the interest of the countryside will always be expendable unless we fight for it. Our only hope is to present a united front, to strengthen our lobby and to put all the power we can into a campaign for educating the townsmen to respect our way of life and to work for and not against our survival. (Barber, 1970, pp.18 - 19)

The delegates to the conference divided into six syndicates, each of which was given a different set of objectives. The first four were agricultural: (a) intensive cereal growing and pig production; (b) intensive arable, but excluding field vegetables; (c) dairy farming; (d) intensive arable, including field vegetables. The two conservation syndicates presented their findings jointly. They had three objectives: first, to maintain as far as possible the diversity of habitat and the richness of wildlife which the farm supported, as far as was consistent with efficient agriculture; second, to preserve the farm's amenity; and third, to provide the farmer with a return from game. Whereas the conservationists were instructed to take into account the needs for efficient farming, none of the agricultural syndicates was asked to consider conservation in the development of its plans. Rather they all were required to develop the land to produce a good income from it, under the burden of a full mortgage and commercial interest rates in order to demonstrate "the need for intensive farming systems to meet current costs of land purchase" (Barber, 1970, p.22).

The delegates became involved in the subject through the syndicates and were confronted with the critical choices to be made when the syndicates' plans were evaluated by conservation specialists. They also had the opportunity to work with others and hear their points of view. The subsequent discussions did, indeed, reveal much common ground as well as a willingness amongst participants to discuss mutual problems and to collaborate in working out possible solutions. There was general agreement over such matters as the need for more tree planting and to retain certain landscape features and interesting habitats on marginal land. There was also much discussion on the best way to manage hedges. However, the use of agricultural pesticides and their impact generated a heated debate, indicating that on certain crucial issues consensus would be difficult to achieve.

The final aim of the conference was to develop a compromise plan, a task which fell to Derek Barber. This involved, in the main, the identification of key features which might be retained even under an intensive agricultural system. The features chosen for retention were dependent on both agricultural considerations and the available data on the distribution of flora and fauna. Attention was also directed towards possibilities for new features, especially woodland planting, and included the suggestion by Barber that an area of rough chalk grassland should be planted. The achievement of compromise was thus essentially a mapping exercise, based on identifying small areas of land which could be retained for conservation objectives without adversely affecting the amount of land available for production. The crucial shortcoming of this approach as it developed from Silsoe onwards was, of course, its inability to take into account how the intensification of agriculture was likely to have increasingly adverse consequences for the remaining wildlife habitat. The message that the *Shooting Times* correspondent took away from the conference – that "provided one preserves the best habitats, it is possible to be quite ruthless with the rest" – was an unfortunate one, therefore (quoted in the Silsoe Report, p.97).

One person at the conference was critical of the failure to present an ecological critique of the various farm improvement proposals. He was Jon Tinker, the environmental correspondent, for *New Scientist*, who, in his report commented:

> There was no mention in the farm plans of what the land might be like in 50 years or even 10 years; there was no concern for maintaining the microflora or physical structure of the soil; there was no discussion of the hazards of nitrate run-off into the streams. Conservationists interested themselves exclusively in the tactical

improvement of farm rationalisation, without criticising the long-term implications. (Tinker, 1969)

Nevertheless, the contemporary costs of wildlife conservation in terms of outlay and lost income, and the profitability of farming at differing intensities were clearly demonstrated. The eventual additional cost of Barber's compromise scheme was worked out at around £1 per acre per annum which was almost equal to the estimated increase in shooting values (and equivalent to one twelfth of the prevailing rent charge). Less intensive farming systems, though, implied reduced profitability. The quantification of such costs sharply identified the conflicts between farming and conservation, and served to illustrate the much more acute pressures to intensify on those farmers who were heavily mortgaged compared with those who were financially secure. Inevitably, this raised the question of who should pay to retain wildlife and landscape features. Subsequently, Barber formulated the issue as follows:

> Looking ahead, do we appeal indefinitely to farmers' altruism? Or do we need far more in the way of injecting non-farm income into on-farm conservation . . . . in the form perhaps of some kind of government grant or a tax concession to those who undertake practical contributions to the public interest? (Barber, 1971)

Three hundred copies of a report on the Silsoe Exercise were issued soon after the event, in July 1969. A year later, in May 1970, an expanded and edited version of the report was published (Barber, 1970). *Farming and Wildlife: a Study in Compromise* provides a detailed account of the deliberations of the meeting. In addition, the concluding summary chapter of the report was issued in leaflet form at the suggestion of Michael Darke of the NFU who felt that few farmers would purchase the full report. Both the report and the leaflet were launched at a press conference held in October 1970 at the Knightsbridge headquarters of the NFU. This and the exercise itself served to awaken the agricultural press to the subject of conservation, and copies of the leaflet were sent to all NFU members.

After Silsoe

Subsequent to the Conference the organizing committee continued to meet to discuss the results, make arrangements for the publication of the proceedings, and consider future developments. It was from this committee that FWAG developed. At the end of October 1969 David Lea circulated a note suggesting the appointment of a full-time officer "to pursue the various promising lines of enquiry and action which arose from the Silsoe Conference". Efforts to secure grant aid for the post were unsuccessful. Lea was not to be deterred though. He persuaded the RSPB Council to fund the appointment jointly with the SPNR, but had no intention of making it an RSPB/SPNR post.

The officer was, instead, to be responsible to the Silsoe Committee, and thus not specifically a representative of either farming or conservation interests. Lea suggested that the person appointed should be seconded from a farming organization, and the subsequent Silsoe Committee meeting strongly urged that the person should be seconded from NAAS. It was clear that a close working relationship with NAAS was envisaged. The imbalance in this is clear, although it did not seem incongruous to those involved at the time. For, whereas the conservation side was to provide the funds and the premises, the agricultural side was to supply the man. In other words, at least initially, it was the conservationists who were wooing the farmers with NAAS arguably acting as honest broker. It was agreed, in consultation with NAAS, that a small team comprising representatives from each of the NFU, SPNR and RSPB should make the appointment. The tasks to be performed by the officer were laid out by Lea and accepted in full by the Committee in January 1970. They were:

(1) To promote liaison with all appropriate authorities on subjects affecting agriculture and wildlife conservation;
(2) To stimulate and assist the organization of further conferences on a local basis;
(3) To approach agricultural colleges and institutes with a view to their including nature conservation on their courses;
(4) To stimulate the investigation of certain specific problems at suitable Research Stations and to disseminate the results of research;
(5) To discuss with Game Advisers how the advice they give could be modified to take account of nature conservation;
(6) To lecture widely to farming and naturalist audiences;
(7) Other tasks might include exploring the possibility of MAFF including conservation exhibits in their displays at agricultural shows.

Thus FWAG was conceived as a ginger group without any executive functions. It was going to persuade farmers, conservationists, the agricultural colleges, the NAAS, among others, to do things. Arguably it could not have started in any other way. To have sought from the start to develop, say, its own advisory service would have been to antagonize existing organizations and would have been inordinately ambitious given its meagre resources. As Derek Barber has explained:

> FWAG's main aim in the early days was to *coax* farmers into conservation thinking .... In 1967, a farmers' view of 'the conservationist' was of some irritating creature, slightly potty, vastly impractical. We needed to 'softly softly' our line to have any hope of success. (Personal Communication, 6 May, 1986)

Lea also proposed that the group should be concerned with wildlife and not general amenity. Although this reflected the specific priorities of the RSPB, there was general agreement on the matter in keeping with the clearer awareness of the harmful impact of modern agriculture on wildlife than on landscape and amenity. Countryside recreation was still conceived of as a problem for farming rather than the other way round. In addition, landscape and recreational conflicts were seen as much more inherently political and intractable, whereas the nature conservationists had promoted, through the 'Countryside in 1970' conferences, a very technocratic notion of conservation, which emphasized, in particular, technical and managerial solutions. This was the approach pursued by FWAG.

At the close of this important meeting of the Silsoe Committee its name was changed to the Farming and Wildlife Advisory Group. Equally, as an afterthought, it was agreed that the Group "might act as a *forum* where any major disagreements between Farming and Nature Conservation could be discussed in a rational way" (FWAG Archives). Thus two of the main ways in which FWAG has subsequently defined itself – first as a forum for national discussion and second as a promoter of county groups and giver of advice – were not strongly emphasized at its inception. Nevertheless, the most important point to make is that the person appointed was to have considerable individual scope for developing the work, with much depending upon his or her energy, skills and contacts. A blow to the plans was NAAS's refusal to offer a senior officer on secondment; and it was then that the Committee decided to advertise the post.

Advertisements were placed in September 1970. The job description, initially offering a two-year contract, contained a more general declaration of the main aim of the post of Farming and Wildlife Adviser:

> To identify the problems of reconciling the needs of modern farming with the conservation of nature, to explore areas of compromise and to make the results of this work as widely known as possible.

The itemized tasks contained one significant addition to the seven points listed above:

> The establishment of a system whereby information could be given to individual farmers about the wildlife on their farm and perhaps advice about its conservation.

The successful candidate would be responsible to FWAG and work under the supervision of David Lea at the RSPB's headquarters at Sandy. The man appointed was Jim Hall, who had recently given up farming on his own account. He had been a tenant of a 100 ha mixed dairying and arable holding in Cambridgeshire where he had been an active NFU member, including Branch Chairman, and an occasional writer on agricultural matters. FWAG was to become closely identified with Hall and vice versa, and it is much to his credit that FWAG made the progress that it did during the 1970s. More than any other person, he established FWAG's identity.

Consequent upon the reorganization of the Silsoe organizing committee the first meeting of the newly formed FWAG was held on 23 April 1970. It comprised the following people: Derek Barber, Eric Carter, Reginald Lofthouse and John Trist (MAFF/NAAS); Michael Darke and Gordon Tickler (NFU); Peter Conder, Ian Ferguson-Lees and David Lea (RSPB); Wilf Dawson and R. Hickling (SPNR); Norman Moore (NC); David Wallace (Cambridge University Department of Land Economy); and Kenneth Williamson (British Trust for Ornithology). The CLA nomination was subsequently taken by L. Gee. The Group was chaired by Barber, who took a strong leadership role in the first few years. He was responsible for preparing background papers on a number of the early farmland exercises, considered below. On a number of occasions his input resolved the kinds of problems inherent to a group comprising representatives from a variety of organizations. Other leading members of the committee included Carter, Moore, Lea and Dawson. The new adviser attended his first meeting in December 1970. Together these people moved FWAG haltingly towards the centre stage of agriculture/conservation politics. This was not an easy task as the FWAG committee did not always have a clear sense of direction and its members had different opinions of what needed to be done. FWAG, was, after all, an *ad hoc* body hoping to evolve into a permanent national organization from a small committee which had been set up to do a specific and limited task. The transition was not without its difficultues and disagreements.

CHAPTER 3
The Development and Internal Politics of National FWAG

Introduction

The Silsoe Conference marked the beginning of a general movement to conserve wildlife on farms. Even so, neither the development of FWAG nor the growth of its influence was without problems. FWAG sought to provide a regular forum for conservation and agricultural interests to meet to discuss matters of mutual interest and concern. In addition, the Group wanted to stimulate the provision of advice and information for farmers keen to incorporate conservation principles into their farm management, and for agricultural advisers and other professionals well positioned to influence farming trends. It was, though, a little known and marginal group with few resources and no formal authority. In the words of one of its founder members, it was "entirely unofficial and with no status" (Carter, 1979). In addition, it experienced certain internal disagreements which reflected both differing appreciations of ecology and wildlife and disagreement about the principles of organization to be followed in furthering FWAG's work.

Nevertheless, by 1986, FWAG was at the head of a network of 64 county groups involving to a greater or lesser extent over 1600 people, most of them prominent in their respective localities in land management, land ownership and conservation. The majority of these groups employed their own full-time farm conservation advisors and an estimated 3000 farmers a year were benefiting from their advisory services. A number of leading rural agencies had, moreover, come to regard FWAG as a central plank in their strategies for promoting the conservation of the farmed landscape and its wildlife. To comprehend how a group initially so marginal to agricultural and conservation politics became so central, we have to understand how it related to the wider institutional and policy context; and how various developments in that context conspired to favour a particular approach to farm conservation problems in official and farming circles, based on voluntary co-operation and compromise – an approach with which FWAG had come to be closely identified.

FWAG's development has crucially depended on the extent to which the organizations from which its members are drawn have been prepared to promote the group. The degree of commitment has varied considerably, with different perceptions of FWAG's role and utility, between organizations and over time. This is not to deny the foresight and initiative of individual FWAG members, but they have perforce had to operate within certain constraints and opportunities created and moulded by the wider institutional context.

Four distinct phases can be discerned in the development of FWAG when particular functions and challenges assumed special prominence in its affairs, as follows:

1970 - 1975	Farming and Wildlife Exercises;
1975 - 1979	The struggle for survival – emphasis on the development of county groups;
1979 - 1983	FWAG in the political limelight;
1983 -	Promotion of a national advisory service for farm conservation.

Although such divisions are inevitably somewhat arbitrary, playing down, as they do, the essential continuity in FWAG's development, they nevertheless correlate with the changing commitments of some of its principal supporting organizations. For example, during the first phase the organizations most actively involved in promoting FWAG were ADAS and the voluntary conservation bodies. The second phase saw some lessening of their commitment but with the NCC stepping into the breach. In the third phase FWAG was firmly embraced and actively promoted by the NFU and the CLA. Finally, in the fourth phase, the Countryside Commission has assumed prominence as FWAG's main sponsor. Each of these developments is reviewed, in turn, below.

Farming and Wildlife Exercises

Initially the FWAG committee devoted much attention to staging more exercises along the lines of Silsoe. This approach had appeal both as a means of bringing conservationists together, and as a way of promulgating the conservation message amongst agriculture's opinion leaders. This was supplemented by the Group's efforts, chiefly through Jim Hall, to publicize the aims of FWAG through articles for the farming press, addressing meetings of farmers, naturalists and others, and seeking to influence the curriculum of agricultural education. Various advisory leaflets were prepared for distribution to farmers, for example, through agricultural shows. A FWAG film on 'Farming and Wildlife' was widely used by ADAS, local conservation groups and commercial firms.

Silsoe provided the impetus for a series of similar exercises and conferences held in different parts of the country (see Table 3.1). Some of these initiatives were independent of FWAG – they attracted some measure of moral and often practical support from the Group. Such

Tabel 3.1   Main Farming and Wildlife Exercises, 1969-1976

National Exercises	Date	Sponsors	Conditions
Silsoe, Beds – on land at Tring, Herts.	1969	Organizing Committee – representative bodies	Lowland – arable and dairy
Dinas, Dyfed	1972	ADAS	Upland – sheep
Kingston Deverill, Wilts.	1973	FWAG, ADAS	Chalk – arable

Local Exercises			
Hammoon, Dorset	1971	ADAS	Lowland – dairy
Churn Estate, Berks.	1972	ADAS, Nature Conservancy, FWAG, Reading University	Chalk – arable
Snelson, Herts.	1973	ADAS, FWAG, Rural Community Council	Lowland – arable
Cowbyers, Northumberland	1974	ADAS, Countryside Commission	Upland – sheep, forestry, game
Sacrewell, Cambs.	1974	ADAS, FWAG, East of England Agricultural Society	Fenland – arable
Letheringham, Suffolk	1974	FWAG	Lowland – arable
Essex	1975	ADAS, Essex County Council	
Reaseheath, Cheshire	1976	College of Agriculture	Lowland – dairy and arable, some wetland
Church Farm, Ely, Cambs.	1976	ADAS	Fenland – arable
Yielden, Beds	1976	FWAG	Lowland – arable

farming and wildlife exercises were major events demanding considerable resources and few occurred on the same scale after 1976. By then FWAG was beginning to turn its limited resources to other pressing matters, such as the setting up of county groups, and the exercise approach was facing criticism from which it never really recovered. Subsequently, local FWAGs have tended to adopt a lower key approach using, in particular, a long-established technique of agricultural advisory work – the farm walk. The preparation for this may or may not involve the compiling of a farm guide. Few, if any, local farm walks involve the formal structuring into syndicates and publication of syndicate findings which were characteristics of the earlier national and regional meetings.

One problem which FWAG faced with regard to the national exercises was determining their appropriate scope. Silsoe was largely confined to wildlife considerations. However, reservations were voiced as to the narrowness of these aims in the discussions over the Dinas Exercise (see pages 113 - 116). Some members felt that the other interests of the hills, forestry, recreation, agricultural problems, should be included; others that such considerations would deviate from the aims of the Group. Agreement was reached that recreational concerns should be excluded. Landscape questions proved more intractable and provoked considerable internal debate. The Countryside Commission was consulted on the handling of landscape issues in preparation for the Chalklands Exercise, but the treatment of these issues was somewhat perfunctory.

In promoting farming and wildlife exercises, particularly with MAFF, FWAG was chary of intruding in any way that might be construed as stifling independent local initiative. This was made clear at a FWAG meeting in October 1970 when Group members were told that some meetings had taken place without their knowledge. It was agreed that MAFF be asked to keep the Group informed about forthcoming conferences. At the same time the Group was keen that a number of key habitat types should be covered in the series. Barber, in particular, pressed for the systematic coverage of major types of land use such as wetlands, uplands and chalklands. There was disagreement within the group, however, over the extent to which its limited resources should be concentrated in this direction.

At a meeting held in November, 1972, there was a discussion of the form of future exercises. Wilf Dawson questioned the cost-effectiveness of the big national events and raised doubts about whether they were the best way of achieving FWAG's objectives. Instead, he suggested a move towards more local demonstrations on farms where conservation was already practised. He saw a need for closer liaison with county trusts as a means of stimulating "advisory capacity at ground level". David Lea endorsed this, claiming "that it had been a great worry to him that the Group had no advisory capacity at the local level." Clearly at this stage in FWAG's development it was felt by some that a new direction was needed or, at the very least, a more direct channelling of Hall's energies. Indeed as Hall records "views have been expressed on the desirability of improving my knowledge of wildlife". To that end, a small steering committee was set up to provide regular guidance to Hall when needed and to look into key issues for the future (see Chapter 4).

Meanwhile preparations went ahead for an uplands exercise in Wales in 1972, a chalkland exercise in Wiltshire in 1973, and a wetlands exercise in Cheshire which, after considerable delays, took place in 1976. The organization of the Wiltshire event was delegated to Barber, Carter and a local representative from each of ADAS and the NCC. The aims were specified in a paper presented to the Group by Barber:

> The exercise should be concentrated on *one basic issue* – the production of blueprints for providing wildlife habitats within the constraints imposed by chalkland arable farming. The remit would require recommendations to be made on the basis of the ability of the farming system to finance the expenditure required and would direct the attention of each syndicate to a particular facet of the reconciliation possibilities – e.g. tree planting on shelter belt and commercial forestry lines, tree planting with shooting in view, restoration of downland etc. (Derek Barber, Proposal for a Prairie Chalkland Exercise, October 1971)

Despite the apparent simplicity of this aim, its execution brought to the surface some of the conflicting viewpoints within the group and the practical limits on the scope for compromise.

The exercise was a thoroughly researched and well-documented affair. It took place in July 1973. In the preceding twelve months a Wildlife Survey Team of over thirty volunteers was active. The team included members of the British Trust for Ornithology who conducted a detailed bird census. The area of Wiltshire covered was 5000 acres (2000 ha) on two estates (see Chapter 7). Some 113 farmers, conservationists and others attended the event and during the three days four syndicates prepared reports, on sport, forestry, wildlife and farming.

From these a 'master plan' was formulated by Peter Dawson, a close colleague of Barber in Gloucestershire ADAS. Some of Dawson's conclusions were not those which have always been associated with the FWAG 'compromise' approach. Much of the exercise was devoted to costing the maintenance of certain wildlife habitats, while allowing modern commercial farming free rein on most of the land. Peter Dawson stated the issue forcefully:

> A fundamental necessity to my mind is the recognition that high farming, which must remain the chief function of the area, and wildlife, are to a large degree incompatible and mutually exclusive and that, accordingly, high farming areas and wildlife areas need to be identified. (MAFF, 1973, p.32)

In other words, the way to reconcile the demands of modern productive farming and wildlife conservation was through effectively zoning them.

This was not necessarily at odds with the FWAG message of progress through compromise. Supporters of FWAG consistently argue that farming with wildlife need not be unprofitable. Rather, profitable farming allows some diversion of resources to wildlife conservation on the remaining marginal areas of a farm, and this was the approach adopted in the Chalklands Exercise. This notion of compromise, however, is often confused with the notion of the compatibility of farming practices and wildlife conservation. That distinction and, by implication, the recognition that stark choices have occasionally to be confronted is often glossed over. Indeed the fudging of the distinction that so often characterizes the compromise message, was evident in the vote of thanks at the close of the Chalkland Exercise given by Michael Stratton, the farmer-chairman of the Wiltshire Trust for Nature Conservation:

> Our first duty as farmers, of course, is to farm as well as we can. And Wiltshire today, I'm proud to say, is wonderfully farmed . . . . we have heard a lot about conservation this weekend, but our first, our pre-eminent, role is to farm well, and good farming is good conservation and is our primary function. (MAFF, 1973, p.45)

The contrast with the earlier remarks by Dawson is striking, but is perhaps indicative not only of a certain ideological confusion but also a terminological one. If Dawson's position is closer to the practical import of a FWAG exercise, so Stratton's sentiment captured more accurately the ideological presentation of the FWAG message. Dawson's remarks clearly caused some consternation among FWAG members, not least because they were the focus of an article in *The Times*, under the headline 'Exercise finds farming and wildlife conservation incompatible' (13 August 1973). Neither FWAG nor its philosophy was mentioned in the piece by the well-known farmer and writer Ralph Whitlock.

FWAG's members, though, were largely preoccupied with the organization of the event and its effectiveness. Kenneth Williamson of the BTO was aggrieved that much voluntary effort had been expended in surveying the wildlife of the farm with little use made of the results. The exercise, he argued, had covered too large an area of land rendering any wildlife survey "unreliable and virtually useless in an analysis which involved appraisal of conservation *vis-à-vis* farm management", ('FWAG and the BTO', Paper circulated to FWAG by K. Williamson, FWAG Archive, 1973). Derek Barber responded that points of fine detail could not be discussed at such a conference and suggested that the BTO had perhaps devoted too much time preparing for the Wiltshire exercise (FWAG Minutes, 1973). Such disagreements pointed up the disparity of information available to farmers and conservationists. The compilation of a great deal of background information was essential for an exercise such as this, but whereas the agricultural data was already available, detailed information in wildlife had to be specially collected.

Williamson was not satisfied, however, and went on to question the value of the national exercise and the direction FWAG was taking:

> It is keen to seek out the big national problems and influence planning and development at the top; and I have the uneasy feeling that if, along the way, something brushes off on the individual small farmer, that is fine. The Group's concept of a worthwhile cause . . . is a big PR exercise on the national scale concerned with more fundamental matters which ultimately will involve political decisions. (Williamson, 'FWAG and the BTO', FWAG Archive, 1973)

Williamson called for more modest exercises aimed at working farmers rather than political leaders, and he berated the group for delegating responsibility for the Reaseheath Wetland Exercise to local bodies in Cheshire. That exercise "would not be unwieldy, the objectives are clearly defined, and the Trust [i.e. the BTO] could entertain every hope of making a positive contribution, capable of being translated into action by farmers and conservationists working in co-operation" (ibid).

A colleague of Williamson at the BTO, Leo Batten, was even more scathing of the Chalklands Exercise:

> We dealt with 5000 acres and nearly £3500 to spend on 'improvements' each year. How many farmers own 5000 acres, and how many have that sort of money to spend on conservation? There seems a real possibility that small farmers may well be overawed by this scale of treatment and give up any idea of conservation because it is far too expensive for them . . . . I cannot help feeling these conferences have outlived their usefulness and should be critically reviewed . . . . the man-hours which went into the preparation of the last conference could have produced more physical effects if they had been used to instruct farmers who seriously wanted ideas relating specifically to their farms. This could best be done by instructing ADAS advisers in conservation matters so that they can give advice to farmers on this subject along with advice on farming

methods. ('Farming and Wildlife Conference' Paper, prepared by L. Batten, FWAG Archive, 1973)

Other members of FWAG did not concur with these sweeping criticisms, which did seem to miss the point that the value of an exercise such as this depended on the careful choice of a farm with potential for the different land uses to be examined. Admittedly, in trying to cover a wide range of problems there was a risk of losing realism and demonstrating compromise for its own sake. Nevertheless Barber reported on the Chalklands Exercise as "an immensely worthwhile and efficiently run public relations exercise". He went on, however, to question its value in a number of crucial respects. Many of the farming faces had been familiar and there had been some "hangers on". He also questioned whether the results justified the immense effort and whether the Silsoe concept had now been fully exploited (FWAG Minutes, July 1973).

There seemed to be general agreement that new approaches were needed and that any future exercise should be more specific and definite in what it was seeking to achieve. Barber suggested a different format for the wetland exercise aimed at such influential bodies as Internal Drainage Boards, River and Local Authorities and possibly selected members of Parliament. Eric Carter reinforced this view arguing that it should be aimed at those who exert pressure on food quality and on pollution, amenity, landscape and agricultural practice. It should, moreover, produce recommendations which would lead to improvements in existing practices. Hall stressed that national exercises were only a small part of FWAG's work and that smaller farmers were catered for by county exercises. FWAG itself could act as a catalyst in promoting these but should not usurp local initiative. In any event the balance of FWAG's effort was shifting and the emphasis on national exercises was being superseded by the increasing commitments at the county level (see Chapter 4).

With a growing sense that the Group was losing momentum in relation to some of its original objectives and was being eclipsed by the efforts of other bodies (particularly the Countryside Commission – see below), a Working Party was set up in April 1976 to review its role and future. This review was also necessitated by FWAG's growing financial difficulties. The Working Party recommended that the Group should continue but with additional members, including representatives from the Countryside Commission, Forestry Commission and Royal Institution of Chartered Surveyors, and with a change in direction. At a meeting of

FWAG that autumn, which marked something of a watershed in its development, it was agreed that:

> The highest priorities of the reformed Group should be the forum function provided by the main committee and the continued establishment of Local Farming and Wildlife Advisory Group Committees and the provision of more support for them . . . . It was felt that the time for further large national demonstration meetings had passed. (FWAG Minutes, 3 November 1976)

The following, contemporary assessment of the FWAG-inspired exercises seems reasonably balanced:

> These conferences have shown that conservation must be part of the plan for the whole farm, or even for a group of farms in a large-scale landscape; that wildlife conservation experts must be prepared to specify exactly what they want; that wildlife conservation has a real cost (estimated at 50p to £1/acre in 1970) and that many farmers and conservationists simply do not understand each other's jargon . . . . Nevertheless the costs have been high, in terms of time and effort spent on organization . . . . The benefits are difficult to assess. Individual delegates must have gained, as did the Ministry of Agriculture officers who were involved. Nevertheless, as introductory educational exercises they have worked well in publicising the principles of conservation and in establishing contacts between farming and conservation groups. But these conferences can only be a base to build upon, not an end in themselves. (Keenleyside, 1977)

The Struggle for Survival: Funding for FWAG

The vigorous pursuit of local groups and publicizing of the FWAG message were often overshadowed by a series of financial and organizational difficulties, which at times assumed almost crisis proportions. At one stage, for example, Hall's contract had only six weeks to run with finance not yet secured for an extension. Time and again the RSPB, itself the contributor of the largest single annual grant, had to step in to balance the books. Though the situation became quite grave in the late 1970s FWAG's existence had been dogged by financial problems from much earlier. Indeed concern and debate over the group's precarious financial position to some extent displaced discussion about its objectives and its direction. Thus in the early and middle years the sheer struggle for survival insulated FWAG from the full impact of some of the major internal arguments which surfaced after 1979.

Initially the bulk of FWAG's funds came from the RSPB and the SPNR. But in 1973 the SPNR declared its intention of reducing its share of funding from the following year. Peter Conder responded that the RSPB might support the whole cost but reasoned that it would strengthen FWAG if other bodies involved – the NFU and CLA, for example – were to make a contribution. Despite this and subsequent appeals, the Group remained heavily dependent for finance upon the RSPB and to a lesser extent the SPNR. It is difficult to avoid the conclusion that the reluctance of the other organizations to provide even modest sums reflected a lack of commitment, notwithstanding the evident commitment of the individuals involved. By 1977, with certain doubts emerging about the value of FWAG within the RSPB, even it began to baulk at shouldering the financial burden. Nevertheless it agreed while alternative funding arrangements were explored, to cover any shortfall until April 1978. Accounts for that year show expenditure of £5870 on the salaries of Hall and a part-time secretary, £1680 on direct expenses, and £3560 on administration and services. The expenditure was covered by grants of £1000 from SPNR, £150 from the NFU and £50 from the College of Estate Management at Reading University, with the lion's share of £9910 provided by the RSPB.

A sub-committee set up under Lofthouse to consider fund raising estimated that £12 000 would be required for the financial year beginning April 1978. Lofthouse pointed out, however, that approaches outside the Group would be easier if it could be stated that the organizations represented on the Group supported it financially. A number of fresh donations from member organizations for the financial year were reported at the first meeting of 1978 of the FWAG Management Committee, but again, with the exception of the RSPB's (£2800), they were all of a very modest nature: College of Estate Management (£100), BTO (£100), NFU (£500), CLA (£500), and SPNR (£500). The RICS declined to give any financial support whereas MAFF stressed the value of its non-monetary contribution – for example, through the assistance ADAS staff gave to county FWAGs – but would not directly fund a voluntary organization. The committee's reaction to MAFF's position was recorded in the Minutes as follows:

> The help which the Group received from MAFF in resources was appreciated by those present, but valuable as that was, the absence of a cash contribution was proving something of a stumbling block in fund raising elsewhere . . . . FWAG was . . . in a state of crisis which only a provision of cash could rectify. (FWAG Management Committee Minutes, 25 January 1978)

The RSPB agreed to provide the necessary additional funding to offer Jim Hall a further contract for the first half of the financial year, and it made another substantial gesture to FWAG by reducing its charges for services. Nevertheless the RSPB was determined to place a limit on its own contributions. FWAG's future clearly lay in the balance and the idea of pursuing charitable status for the Group was shelved as it would only be relevant if the Group was assured of a long-term future.

As the year advanced news of grant applications that had been made to the Countryside Commission and the NCC was anxiously awaited. In the event a grant of £2500 from the NCC saw it through the remainder of the financial year. It was not until the following year, in June 1979, that Keith Turner of the Countryside Commission was able to give an undertaking that the Commission intended to contribute "as soon as money was available" and "in consultation with other statutory bodies" (Letter, FWAG Archive). However, the NCC had already agreed to double its contribution for that year, and with grants promised for the next 2 to 3 years by the W.A. Cadbury Trust and the World Wildlife Fund, prospects were greatly improved. In the autumn of 1979 the Management Committee pronounced FWAG to be solvent for at least the following two years. However by the Spring it became apparent that there would be a shortfall of £1000 for the next financial year. Again aid was sought from the Countryside Commission, this time successfully.

Although FWAG had overcome its previously dire financial situation, its funding was still not on a secure, long-term basis. The bulk came from one-off or short-term grants. Even the sponsorship of its member organizations was not assured. Some were less than prompt in paying what had been promised: by 1978, for instance, the NFU was two years in arrears. Some reconsidered their contributions annually. Others kept them static during this period of rapid inflation, whereas the CLA having doubled its grant to £1000 for 1978-79, dropped back to £500 in 1980. Inexorably, though, the demands on FWAG's budget were increasing: particularly with the growth in the number of county groups.

Hall's impending retirement on reaching the age of 65 in 1981 was a cause of more heart searching as to FWAG's future. In a paper prepared for the May 1980 meeting of the Management Committee, Lofthouse expressed the view that Hall was irreplaceable, both in terms of his experience and the available finance. Several frustrating years as FWAG fund-raiser led him to suggest that FWAG should continue without a full-time adviser. Referring to the dictum that wholly voluntary organizations survive best, Lofthouse proposed the keeping of "minimum records", and seemed to envisage a slimmed-down FWAG forum and almost complete autonomy for the county groups. There was some support for this position particularly within the RSPB, but Hall seemed to express the view of the majority of FWAG members when he reasoned that the 800 or so individuals involved in FWAG up and down the country deserved a better deal than that. In the event, the question was pre-empted by Eric Carter's offer made to the Management Committee meeting of 13 March 1981, and accepted immediately, to take over as full-time adviser. Carter, himself, had just retired as Deputy Director General of ADAS.

The decision was also made that a full-time secretary should be appointed to service national FWAG and the Adviser's work. A sum of £20 000 was estimated as the requirement for 1981-82, and Carter started his time as Adviser with a six-month contract. Appeals to the Countryside Commission, NCC and MAFF were renewed. A request to Peter Walker, the Minister of Agriculture, under the names of some of the chairmen of FWAG's member organizatons, brought once again a rejection which stressed the value of MAFF's input of non-monetary resources. The NCC, though, continued its support and increased its backing such that by 1982 its grant covered almost a third of the group's expenditure. However, the long-term funding of the national body, though still unresolved, was increasingly overshadowed by the question of external support to develop the function of the county groups. As the conflict between agriculture and conservation attracted heightened political interest, so increasingly was the network of county FWAGs recognized in official quarters and agricultural circles as offering a possible foundation for remedial action.

FWAG in the Political Limelight

By the mid 1970s considerable evidence about the adverse impact of modern agricultural technology on wildlife habitats had accumulated. Of the organizations associated with FWAG it was the NCC which took the lead in drawing together the relevant findings. For some years the Nature Conservancy had been investigating the harmful effects of agricultural practices, with much of the work done by its Toxic Chemicals and Wildlife Section under the direction of Norman Moore. To communicate the findings to those responsible for the management of the farmed countryside, an Agricultural Habitat Liaison Group was established in 1971 with members drawn from the Research and Conservation Branches of the Conservancy and a senior officer from MAFF and, later, one from the Department of Agriculture and Fisheries for Scotland. Over the next two years the group met regularly to review and advise its member organizations on research, training programmes, advisory work and practical action. This liaison with the agricultural departments had just been placed on a firm footing when in 1973 the Conservancy was reconstituted, with its former conservation and research functions being split between the new Nature Conservancy Council and Institute of Terrestrial Ecology.

The NCC was thus propelled from being a component of a research council (the National Environment Research Council) to the status of a statutory body with responsibility for promoting nature conservation and with direct access to Ministers. This gave it a more central policy advisory role and one increasingly orientated to change in the wider countryside rather than just being limited to the protection of special sites. Norman Moore was made the NCC's chief advisory officer and it was hardly surprising that, in seeking to develop long-term policies for nature conservation, the NCC should focus initially on the effects of agricultural change.

Moore was assigned to prepare a discussion paper on the issue, and this drew together a range of survey evidence on habitats and individual species indicating unprecedented decline. He showed that there were no easy solutions: nature reserves alone could not sustain the beleaguered fauna and flora. At an early stage, before they were published, Moore discussed his conclusions with the FWAG Committee. He had come to the judgement, he told them, that "modern agriculture was unlikely to produce good conservation practice" and questioned "whether any thing was being achieved through promoting liaison and advice". Various surveys, he suggested, had shown that "awareness was not getting through to most farmers, or being expressed in practice" (FWAG Minutes, 10 July 1975).

Moore's remarks provoked a heated discussion concerning both the pessimism of his conclusions and their implications for FWAG. The response of Michael Darke of the NFU is of particular interest, especially in the light of subsequent developments. Darke recalled his previous suggestion, that the NFU "could usefully compile a list of farmers and landowners who professed to sympathise with the conservation of wildlife" (ibid). This was in keeping with a traditional stance of the NFU: that interest in wildlife was perhaps a suitable minority avocation for large landowners and gentlemen farmers but was undoubtedly a distraction for the working farmer. It was a stance that was very soon to be completely abandoned as too complacent in the face of the gathering political momentum of the conservationist critique of modern agriculture.

Moore's strictures caused consternation within FWAG and at the next meeting, in his absence, concern was expressed lest his study "challenge the value of FWAG's work, its tactics and the practices it was advocating" (FWAG Minutes, 12 November 1975). It was decided that the Chairman should write to the Director of the NCC asking that the results of Moore's deliberations not be made public until there had been consultation on his findings. Nevertheless, Moore's work led the NCC to conclude:

> Only if the nation specifically plans for wildlife will it survive to any significant extent in the lowlands and more fertile uplands of Britain. All too frequently the continued existence of the most valuable sites for nature conservation is the result of the perpetuation of some special farming practice which cannot necessarily be expected to persist much longer. (NCC, 1976, p.13)

In its subsequent report, *Nature Conservation and Agriculture* (1977), the NCC recommended safeguards for SSSIs against adverse agricultural operations and attention to the conservation of the remaining semi-natural areas on farmland through the information and grants offered to farmers. It called for more accessible and more directly usable advice on conservation and suggested that specially trained 'Rural Advisers' might be the way to maintain direct contact with farmers and landowners.

This and other studies, including the Countryside Commission's influential *New Agricultural Landscapes* (1974), established the parameters for a wide-ranging debate, that gathered momentum during the following years, on how the forces intensifying production might be moderated in the interests of conservation (see, for example, Davidson and Lloyd 1977, Shoard 1980, Green 1981). With the growing appreciation of the totality of agricultural change, pressures built up from conservation groups to regulate the environmental impact of agricultural development. The farming lobby, not unnaturally, resisted any such curbs. The CLA and the NFU meanwhile responded to the charges of conservationists by drawing upon time-honoured ideological resources to present farmers as stewards of the countryside. They also laid great stress on the need to retain the goodwill and voluntary co-operation of the farming community if practical remedies were to be found to conservation problems.

But, in addition, their defensive arguments against any form of compulsion were increasingly characterized by the prominence given to their support for FWAG. It was, at first, presented as an example of the co-operative spirit at work but was subsequently highlighted as a pattern of the way the goodwill of the farming community should constructively be mobilized. Previously the NFU and the CLA had done little actively to promote FWAG. Eric Carter, for one, had been led to complain that "The NFU and the CLA should be more energetic in their support" and had urged them to "carry out a public relations campaign on behalf of FWAG" (Minutes, FWAG Chairman's Working Party, 21 May 1976). There is, indeed, a stark contrast in their publicity between the high profile accorded FWAG in the late seventies and early eighties and the sparse references to it in the mid-seventies. In the *British Farmer and Stockbreeder*, the official magazine of the NFU, for example, there were only 3 articles between 1973 and 1977 that gave any coverage to FWAG. No wonder, then, that an article of October 1978 in the NFU's other organ, *Insight*, explaining the Group's role could open with the words, "Although it has now been in operation for about 10 years many farmers may never have heard of FWAG and will have no idea of its aims and objectives". In contrast, between 1978 and 1982, FWAG was covered in some 15 articles in *British Farmer and Stockbreeder* including major and leading features.

An interview survey of farmers' attitudes to wildlife conservation conducted by the Ministry of Agriculture in 1975 gave some indication of the likely scale of demand for wildlife advice amongst farmers (MAFF, 1976). An equal proportion – about 13 per cent – of those interviewed intended to reduce some of the 'non-farmed' areas of their land during the following ten years as did those who intended to improve existing wildlife habitats or create new ones. The areas most likely to be reduced included unimproved grazing, scrub, quarries and wet and marshy land, whereas the habitats most likely to be improved or created were woods, hedges and wetlands. Positive conservation intentions were correlated with personal interest in wildlife or game, membership of conservation bodies and ownership of large farms. Nevertheless, 9 out of 10 farmers expressed an interest in allowing at least some types of wildlife to exist on their farms, though it is unclear whether many realized what this might entail. One-quarter of the farmers admitted virtually no knowledge of wildlife. All in all, therefore, there seemed considerable potential for advice on wildlife conservation, with 60 per cent of those interviewed saying that they would welcome such advice, including 26 per cent who wanted both written advice and a visit by a wildlife specialist.

The CLA and NFU turned their attention jointly to the question of conservation guidance in a leaflet published in 1977, entitled *Caring for the Countryside*. This statement of intent acknowledged that farmers and landowners have an important part to play in the conservation of scenery and wildlife; and it presented a basic conservation guide for their members, with pointers on the maintenance of desirable semi-natural habitats on farms and the creation of new landscape features. Encouragement was given to seek appropriate advice when necessary:

> The local ADAS officer is the first contact for advice. The regional officers of the Countryside Commission, Nature Conservancy Council and Forestry Commission together with local authorities will often be able to help with both advice and grants. There are also a wide range of voluntary bodies including local conservation or naturalists' trusts and farming and wildlife groups who are ready and willing to help. (*Caring for the Countryside*)

This joint statement (which three members of national FWAG had helped prepare) was very much a response to overtures seeking an accommodation between conservation and agriculture both from the Countryside Commission following its *New Agricultural Landscapes* study, and from the NCC following its *Nature Conservation and Agriculture* report. Significantly, the Countryside Review Committee, on which both bodies were represented, along with senior civil servants from the DoE and MAFF, declared in its paper, *Food Production in the Countryside* (1978) that "The goodwill inherent in this statement should be one of the cornerstones of future government policy for conservation". On the premise that "any new policy must have the broad support of the farming community", it roundly rejected the imposition of controls over agriculture, arguing that they would be cumbersome, costly and unconstructive. Instead, it called – with words which were surely mellifluous to the CLA and NFU – for a "voluntary and flexible policy, based on advice, encouragement, education and financial inducements". The rearguard action had successfully denied those seeking fundamental reform any sort of bridgehead in official circles. The CLA and NFU continued to work to consolidate this advantage and increasingly they invoked their support for FWAG as an embodiment of the farming community's wider commitment to voluntary co-operation.

In a 1978 issue of the NFU's journal *Insight* emphasis is placed on FWAG in an article entitled 'Wildlife and the Farmer':

> If the aims set out in *Caring for the Countryside* are to be achieved it will not be by the posturing and strident tones that have been forthcoming from certain elements of the conservation lobby, nor by compulsion and the big stick. A much more positive and long-lasting contribution is likely to be made by the type of co-operative and practical approach which is the aim of FWAG, encouraging understanding, compromise by all parties and local involvement. Here, surely, is one positive way of moving forward and it is to be hoped that farmers will support and participate in the work of groups on an increasing scale. (Sly, 1978, p.8)

In view of this it is perhaps surprising that no specific mention was made of FWAG in *Caring for the Countryside* – an omission which did not escape Jim Hall's notice. In a letter to the CLA (10 April 1978) Hall drew parallels between the use of the word "custodian" to characterize the farmer's role in *Caring for the Countryside* and by farming leaders in the 'Countryside in 1970' conferences. In the latter case, it had remained just "a resounding claim", but this time, Hall urged, it must be followed through into practice. In a follow-up letter Hall appealed for a "more positive contribution" by CLA members to the county FWAG committees, explaining that "we have had to overcome some suspicions of our motives, and in some cases, . . . a complacency about the situation as it is, and a bland assumption that the owner always knows best" (Hall to William de Salis, 11 April 1978).

In subsequent issues of *Country Landowner* (the CLA's monthly magazine), *Insight*, *British Farmer and Stockbreeder* and NFU county magazines there appeared more and more news items and articles about FWAG, but still the NFU and CLA continued to regard it as one forum among many for discussing rural problems and to look to an expanded remit for ADAS as the main source of future conservation advice for farmers. In this respect, much was expected – not only by the agricultural interest groups but by the statutory conservation agencies also – of the outcome of a review initiated in May 1977 by MAFF's Advisory Council for Agriculture and Horticulture. The Council, under the chairmanship of Sir Nigel Strutt, was asked "To advise on ways in which the Ministry could best contribute towards reconciling the national requirement for economic agricultural and horticultural production with the development of other national objectives in the countryside in the light of public interest in recreation and access and in conservation and amenity".

The Council took evidence from a wide range of official and voluntary countryside organizations. In its report of 1978 it recorded "a virtually unanimous view that MAFF, which had tended generally to adopt a low profile in many matters affecting conservation, should now assume a much wider, and more openly committed, role" (para. 96). Endorsing this view, the Council suggested that the most important single need was greater provision of more expert advice and it identified ADAS as "uniquely fitted" to undertake this:

> It has close contact with the industry and has the confidence of farmers . . . it is the organisation best capable of expeditiously promoting conservation across a wide spectrum of farmers and landowners. At farm level, with these new responsibilities, ADAS would be able to integrate its conservation advice with that on technical and economic matters. (para. 182)

This,it recognized, would entail ADAS developing its own expertise in conservation matters as well as working more closely with agencies such as the NCC and the Countryside Commission. A wider responsibility for MAFF would also confer a consequent obligation to include conservation criteria in the grant-aid schemes it administered.

Hopes for reform along these lines received a setback with the change of government in 1979 and the advent of a Conservative government committed both to cutting the civil service and to diminishing the burden of regulatory constraints on business and industry. Rather than expand the remit of ADAS, Ministers looked for significant staff cuts. In 1980, after a review of the service by Sir Derek Rayner in pursuit of administrative economies, the main strategic opportunity for ADAS staff to have intervened to raise conservation factors in farm decision-making was erased with the removal of the requirement that farmers should seek prior approval to carry out changes for which they intended to claim grant.

At least initially, the new Government also showed a marked antipathy towards 'quangos', not least in the environmental field (Lowe and Flynn, 1989). Rather than turn to the Government's statutory advisers on conservation or other official sources of advice, Ministers sought the counsel of the agricultural interest groups to agree an approach in which self-regulation within farming would be relied upon to solve conservation problems. Within days of the Conservatives taking office, both the CLA and NFU had separate meetings with agriculture and environment ministers to discuss their legislative proposals. Broad agreement on a Wildlife and Countryside Bill was reached and from this point through to the enactment of the legislation in October 1981 the Government, the CLA and the NFU remained in essential accord on the philosophy of the Bill – namely, that conservation objectives should be secured through the voluntary co-operation of farmers and landowners, encouraged where necessary by management agreements drafted and financed by conservation agencies or planning authorities (Lowe et al., 1986).

The absence of significant safeguards for important scenic and wildlife resources provoked opposition from conservation and amenity organizations. Some groups not represented on FWAG, including Friends of the Earth and the Ramblers' Association, took up the call for planning controls to be extended to agricultural development. Other organizations that were represented, including the RSPB, the SPNR (now RSNC) and the Countryside Commission, while stopping short of endorsing full-blown planning controls, did see a vital necessity for back-up powers to management agreements to prevent the destruction of valued landscapes and habitats. As the debate surrounding the Bill gathered pace in Parliament and the media, therefore, the agriculture and conservation interests found themselves on the opposite sides of an increasingly polarized debate which tended, if anything, to promulgate and reinforce stereotypes concerning 'extremist conservationists' and 'rapacious farmers'. Gloucestershire FWAG (1982), for example, complained that the publication of Marion Shoard's *The Theft of the Countryside* and the extensive media coverage it had received had antagonized many farmers and made for some difficulties in encouraging practical conservation projects. Controversy over the legislation, though, did not end with the passage of the Act. Whereas the CLA, the NFU and the Government were committed to its success, conservationists remained sceptical over the procedures and resources for its implementation and the monitoring of its impact.

National FWAG maintained its cohesion through this difficult period largely by avoiding debate on the matter. As the Wildlife and Countryside Bill completed its parliamentary passage, a discussion was held on its implications for FWAG. Jim Hall reported that he had found "the trust and respect built in to county FWAGs between farmers, landowners and conservation members to be standing the strain" (FWAG Minutes, 15 July

1981). The general hope of the meeting was that FWAG would "continue to preach conciliation which was most important when there was an alarming risk attached to present polarisation in reaction to the sharpness of the debate" (ibid.). However, the whole terrain of conservation politics had changed and inevitably this had major implications for FWAG's role.

An important shift in attitudes towards FWAG occurred amongst its member organizations during the passage and early implementation of the *Wildlife and Countryside Act*. When pressed to justify their claim that voluntary co-operation would be sufficient to safeguard rural habitats and landscapes in the face of widespread evidence that these were being destroyed by modern farming practices, agricultural leaders and ministers publicly associated themselves ever more closely with FWAG as a tangible expression of the principle of voluntary co-operation. Indeed, when challenged on the apparent failings of some aspect of the Act and evidence of continued conservation losses, Ministers almost routinely referred to FWAG as indicating a more constructive attitude at work within the farming and landowning community. The House of Lords, for example, debated the workings of the Act on 13 June 1984. The NFU's briefing paper for peers reiterated the Union's established stance that "ultimately the long-term diversity and interests of the countryside can only be maintained and enhanced through co-operation and voluntary means"; and then added, by way of elaboration, that the Union had "committed itself to a variety of initiatives aimed at furthering the 'conservation ethic' within agriculture, for example through support in cash and kind for increasingly important Farming and Wildlife Advisory Groups."

Nevertheless, the debate raised many criticisms of the shortcomings of the Act; but Lord Skelmersdale, replying for the Government, commented, echoing the sentiments of the NFU:

> I believe that there is a genuine will among everyone involved to break the log jam between the needs of conservation and of agriculture. As well as persuading farmers to be more aware of what they can do for conservation, we need to do more to educate conservationists in the requirements of the farmer. A great advancement in this respect . . . are the inelegantly named FWAGs . . . . The formation of these FWAGs is an expression of confidence in the future. (House of Lords Debates, 13 June 1984, c.1245)

As the agricultural interests and ministers more warmly embraced FWAG, so the conservation interests reassessed their support for it. The voluntary conservation bodies moved to disengage themselves somewhat, particularly from the role of FWAG as a consensus-forming body. As early as 1976 John Andrews, the Head of the Conservation Planning Department of the RSPB, had anticipated some of the difficulties that might arise for that organization as FWAG developed a corporate identity and became associated with a distinctive 'middle-way' approach to conservation:

> When it was established, FWAG provided a forum where apparently conflicting interests could come together, air their views and work towards a compromise . . . . The Group had particular strength in that it was neither a farming group nor a wildlife group . . . . My impression is that its membership is now so strongly weighted towards people with a sympathy for and understanding of wildlife problems that it is no longer bringing together opposing sides on a neutral ground where they can seek compromise. If this is so it will also lose credibility in the eyes of some of the farming community and become just another wildlife body. It will also mean that, unless they alter their present policies, neither the RSPB nor the SPNR will have an effective input to the major areas of conflict with agriculture. I have the feeling that FWAG is developing a sense of corporate identity and a desire to promote itself as an organisation. Again I think that this is wrong. It is simply one of a number of methods by which opposing interests can come together and perhaps generate new ideas . . . . Similarly, I do not see how FWAG can represent a view. There are few common views between wildlife and agricultural interests and it is necessary for the individual bodies to express their own opinions . . . . In summary, I think that the RSPB should itself look at a number of agricultural problems and take direct action on them, that FWAG should be one medium through which action could be pursued by providing a forum for opposing interests, that FWAG's growing identification with the wildlife lobby should be stopped (it should be excluded from RSPB annual reports where it appears to be yet another department), and that the Adviser should move into the realm of gathering data on problems that require study and away from attempting to represent a "FWAG view" which I think is for the moment at least a mythical beast. (John Andrews to John Parslow, internal RSPB memo, January 1976)

It is not clear how widely shared was Andrews's assessment within the RSPB, but from this time the organization began to put limits on its previously strong and unquestioned support for FWAG. The late 1970s was a period when the RSPB deliberately sought to extend its political influence in keeping with its emergence as the largest wildlife organization enjoying mass support. It was naturally unwilling for its links with the agricultural departments to be circumscribed by its association with FWAG, which no longer seemed to offer an effective bridgehead into agricultural policy making, if it ever had. FWAG's minimal involvement in the Wildlife and Countryside Bill confirmed its marginality to serious policy negotiations. Active MAFF participation in FWAG was evidently no substitute for consultative status with the Ministry, and a suspicion arose lest the Ministry's enthusiasm for FWAG was simply a sop to conservation interests. The Director of the RSPB was led openly to express his disillusionment:

> Rather than seeking to change farming policies and practice, we chose to become a prime mover of the Farming and Wildlife Group [which] sought to explore the opportunities for compromise between production and conservation on the farm. Latterly, however, it has become clear that collaboration is not enough – the tide of agricultural change runs so strongly that it threatens to overwhelm some farmland birds. But to alter agricultural policy is no easy matter. (*Birds*, Summer 1984)

Conservation groups, whether or not involved in FWAG, were anxious that the rhetoric surrounding it should not deflect attention from the need for policy reform. As the Director of the CPRE argued:

> The work of the Farming and Wildlife Advisory Groups, ADAS and the thousands of individual farmers already committed to conservation will only succeed if there are major changes . . . to the financial incentives which lie at the root of agricultural policy. (CPRE Press Release, 9 March 1984)

The statutory conservation agencies adopted a different general stance towards FWAG. Whatever the opinion held by their senior staff about FWAG's utility – and the Countryside Commission's representative on FWAG expressed his concern that "MAFF is tending to use the organisation as a cosmetic screen, which allows them to parade conservationist ideals without having to do much about it" (Countryside Commission internal memo, March 1980) – it was abundantly clear that voluntary co-operation was the central political principle of the Government's conservation strategy and they had no option but to toe the line. The Countryside Commission, for example, had evinced little regard for FWAG but became its main financial backer after the appointment of Derek Barber as Chairman of the Commission in 1981 (see next section). The voluntary and statutory conservation bodies, nevertheless, concurred in wanting to see FWAG given a narrower but sharper and more practical role, so as to enhance its value as a positive force for conservation of farmland habitats and landscapes. Intense lobbying had won important safeguards for designated sites, but in the wider countryside the only surety for wildlife and scenery was indeed the goodwill and commitment of farmers, individually and collectively. FWAG offered a potential means whereby such commitment might be engendered and nourished. As the Director of the Countryside Commission, Adrian Phillips, commented:

> Management agreements are only able to deal with relatively small areas of exceptional countryside. Elsewhere, conservation must continue to depend very largely upon the goodwill of farmers. Hence the importance which the Countryside Commission and the Nature Conservancy Council attach to the work of the Farming and Wildlife Advisory Groups. Their role will surely expand over the next few years. (Country Commission Information Bulletin No.60, 1983)

There was also certainly a keenness to see some of the moral and rhetorical resources that the agricultural lobby had so demonstratively invested in FWAG and the voluntary principle converted into substantive support and action for conservation. As Jim Hall commented in an appeal to the CLA for a more active contribution to FWAGs: "Your members tend to take a passive, rather 'stand back' role, whereas I would like to see them not only pursuing a public relations exercise . . . but be much more constructive than they are"(Jim Hall to James Douglas, 26 February 1981). Norman Moore also pressed strongly for much clearer and more rigorous thinking and guidance within FWAG concerning conservation priorities, particularly the much greater value of retaining existing habitats compared with creating new ones. Otherwise, he argued, "There is a danger that people think that cosmetic treatment is all we are asking them to do. There is a real danger of us trivialising the whole exercise" (FWAG Minutes, 6 March 1980).

These pressures from the conservation side coincided with the imperative need, in the wake of the *Wildlife and Countryside Act*, for agricultural leaders to prove that the voluntary principle, with all the claims that had been made for it, could work in practice and indeed would be equal to the critical role to which it had been assigned. By raising expectations and heightening awareness of the issues, the passing of the Act resulted in a marked shift in the terms of the conservation debate. Whereas previously the onus had been on the conservationists to demonstrate the harmful effects of changing farming and forestry practices, the onus was now on the agricultural community and the Government to demonstrate that the Act was being effective in halting the destruction of wildlife habitats and landscapes. With a will born of necessity, the NFU and the CLA embarked on the task of creating a consensus around the Act. Understandably, they were concerned to ensure that the stewardship practices of their members more than matched the rhetoric of the voluntary case. The main vehicle they chose in their efforts to highlight and reinforce a conservation ethic and co-operative spirit amongst farmers and landowners was FWAG. They too, therefore, sought to give FWAG a more effective operational role in the self-regulation of the agricultural community. Their confidence in its ability to fulfil this role was enhanced by the appointment as national FWAG adviser of Eric Carter following his retirement as Deputy Director General of ADAS.

Efforts ensued to extend and, where necessary, revive the network of county FWAGs and to develop their capacity to provide conservation advice (see Chapter 4). NFU and CLA headquarters promoted this drive which met with enthusiastic support in most areas. A CLA booklet, *Planning and the Countryside* (1982), commended FWAGs in the following terms:

> Their purpose is to provide realistic and practical advice to farmers and owners on ways in which modern farming can be carried on with due regard to wildlife and landscape features. FWAGs are made up of members of the official bodies concerned with the countryside and its conservation and representatives of the voluntary conservation bodies and of the owners and farmers. Advice from this source is therefore respected as being independent and balanced.

A survey of CLA members found that more than a quarter were in touch with their local FWAGs (CLA, 1982). The NFU's annual report for 1983, meanwhile, described FWAG as "an essential means of providing practical advice to the farming community", and Fred Elliott, Chairman of the NFU Parliamentary Committee, pointed out to the Union's annual meeting what was at stake if the membership failed adequately to respond:

> We must become even more conscious of our environment. If we do not take action ourselves then we face the threat of greater controls over agriculture, however impractical they may be. I urge all farmers to make conservation a part of their everyday farming . . . . We will continue to press for realistic policies and practical measures to assist farmers to work in an environmentally acceptable manner – but we must have the support of farmers on the ground if we are to achieve this. One of the best ways of demonstrating a commitment is by supporting organisations such as the Farming and Wildlife Advisory Groups. (NFU Press Release, 14 February 1984)

If county FWAGs were to play a self-regulatory role, however, it was important they did more than preach to the converted. National FWAG examined how to build conservation into the agricultural curriculum. In 1981, under its auspices and through the efforts particularly of Jim Hall and Graham Suggett, the Principal of Warwickshire College of Agriculture, the Conservation in Agricultural Education Guidance Group was set up. Adopting an equally pragmatic perspective to that of its parent body, the CAEGG deliberately chose to avoid any approach to curriculum change that might be construed as "mandatory or prescriptive". Instead, it set about preparing notes for distribution to agricultural lecturers that could be slotted into their existing courses (see pages 161-3). A basic assumption in drawing up the notes, on such topics as insecticides, straw and stubble burning, land reclamation and aerial spraying, was that each was "an integral and necessary part of modern farming and the leaflets therefore merely seek to show the possible environmental effects of the practice" (Suggett 1982).

While CAEGG sought to sensitize future generations of farmers, the immediate need was for widely available advice for existing farmers and landowners. It was this call for professional extension staff which, once again, raised the question of the funding of FWAG. As the CLA (1982) commented, "While firmly believing that [the] voluntary approach is the best one the CLA has urged that the system is backed up with sufficient money and able, practical advisers so it can run smoothly and effectively".

Promotion of a National Advisory Service for Farm Conservation: FWAG and the Countryside Commission

In February 1984, in what represented a major triumph for FWAG, a Farming and Wildlife Trust was launched principally to raise the money needed to fund the employment of 'farm conservation advisers' in the counties. It was planned to appoint up to 30 advisers over the following 2 to 3 years and to this end an appeal target of £500 000 was established. Significantly, half of this money had already been pledged, in advance of the formal launch, and by far the biggest sponsor was the Countryside Commission which, over the next ten years, was prepared to commit up to one million pounds to the venture (Countryside Commission, Annual Report, 1983-84, p.7). The transformation in FWAG's fortunes to a point where it was poised to become the lead agency for conservation advice to farmers caused Derek Barber to remark that "all has changed to an extent that the founder members of FWAG still find it almost unbelievable" (Countryside Commission press release, 15 February 1984).

Equally remarkable was the change in the relationship between FWAG and the Countryside Commission which underpinned FWAG's new-found fortune and which, more than anyone else, Barber had helped to engineer. For the first ten years of their existence the relationship between FWAG and the Countryside Commission had been marked by indifference and suspicion. There had been a number of obstacles to a more harmonious relationship. First, FWAG had originally excluded landscape and amenity issues as beyond its scope and, though the distinction between nature and landscape conservation proved impossible to sustain, the expediency of inviting representation from the Countryside Commission was at an early stage rejected on just these grounds (FWAG Minutes, 10 May 1971). This decision was singularly unfortunate and cannot be completely divorced from the damaging institutional division and rivalry that then prevailed between landscape and conservation interests. At the time the Commission was beginning to take advantage of its new research and experimentation powers, and the country-wide responsibilities conferred by the 1968 *Countryside Act*, to promote the concept of countryside management. This approach to local land-use conflicts was based on the principle that, in the absence of planning controls over farming and forestry activities, public agencies should seek to secure conservation or recreation aims on private land, in co-operation with the owner, through advice, project work or payments designed to modify management practices. Such an accommodative approach would seem to have a close affinity with the conciliatory outlook promoted by FWAG. The Commission pursued and refined this approach through experimental projects in various selected areas: in heritage coasts, in the uplands, on the urban fringe, and in the intensively farmed lowlands. Inevitably given its vastly superior resources, its efforts completely overshadowed those of FWAG, and, arguably, the Commission's exclusion from the Group contributed more than any other factor to the latter's marginality during the early and mid-1970s.

Perhaps the area in which there was most overlap was the Commission's farm interpretation work and the series of projects and initiatives following up the *New Agricultural Landscapes* (NAL) study. In the former, the Commission explored the potential of farm open days, trails and interpretation centres for increasing the public's knowledge of farming processes as well as demonstrating appropriate techniques to farmers. The very same processes were transforming much of the lowland countryside as the *New Agricultural Landscapes* study (1974) brought home, in demonstrating that changes in farming could comprehensively alter 'the landscape' rather than just individual features of it. The Commission was optimistic, however, that with proper guidance, modern agriculture could produce a new, though equally attractive countryside, largely through planting or conserving the bits the 'high-tech' farmers had no use for.

To follow up this study, therefore, between 1974 and 1978 the Commission established a series of Demonstration Farms – 11 in all – to show how, without high capital and management investment, conservation interests might be combined with economic farming. Highly commercial farms, representative of a diversity of regional and landscape types, were selected on which to carry out and test landscape and wildlife conservation techniques compatible with modern methods of agriculture; and then to demonstrate the results to farmers, landowners and agricultural advisers. This was a logical extension of the FWAG-exercise approach, producing not only a farm-conservation plan but implementing it also. The long-term aim, as the Commission explained, was "to link these demonstration farms to a conservation advisory unit jointly staffed by the Countryside Commission, MAFF and the Nature Conservancy Council" (Annual Report, 1974-5, p.18). The Commission also established five experimental countryside management projects in

Suffolk, Hereford and Worcester, Cambridgeshire, Bedfordshire and Leicestershire with the aim of improving the landscape over substantial tracts of countryside. A project officer with a small budget was employed by the local authority to encourage and assist farmers and landowners in carrying out conservation projects on their land.

It was perhaps inevitable that a certain degree of rivalry and tension should develop between the Commission and FWAG. FWAG responded to the Commission's consultation exercise following *New Agricultural Landscapes* with a complaint that it had failed to recognize the work which FWAG had already done in this field. Then, when the Commission's Demonstration Farms Project was being set up, a correspondent in *British Farmer and Stockbreeder* accused the Commission of attempting a takeover of FWAG (10 April and 22 May 1976). Within FWAG hopes were kindled of some involvement in the disbursement of the £50 000 allocated by the Commission for the establishment of demonstration farms, but the only concession won was for the Commission not to set up a demonstration farm in an area without consulting the local FWAG where one existed, in return for a commitment not to set up local FWAGs where they would overlap with the Commission's project (see page 71). The situation was not helped by personality conflicts. Reg Hookway, Director of the Countryside Commission, was a forthright person who did not mind upsetting vested interests, not least within agriculture. For his part, Jim Hall seemed to share an attitude then prevalent within the farming industry which portrayed the Countryside Commission as an outside and interfering element. Certainly, his approaches to the Commission tended to be abrasive and uncompromising, and served largely to antagonize.

As the Commission extended its interests in the farmed countryside – including a new grant scheme to promote tree planting and woodland management – it became clear that FWAG could no longer afford to exclude it. Moreover, in 1976 the Commission began to offer grants to voluntary bodies active in its field of interest to help them extend their role and strengthen their organization. The Woodland Trust was one early beneficiary and Commission funding facilitated its transformation from a small, local band of enthusiasts into an influential national organization with its own staff. By this time, in contrast, the SPNR had greatly reduced its contributions to FWAG, and the RSPB was looking to diminish its own financial commitment. A period of chronic financial uncertainty for FWAG was just beginning. The prospect of the Countryside Commission filling the gap was understandably tempting, therefore, and from 1977 onwards FWAG repeatedly approached the Commission for grant aid. In

December 1976 FWAG Chairman, Ian Prestt, invited the Countryside Commission to become a supporter of FWAG. The invitation arose as a consequence of the review of FWAG's role which had included discussions between Prestt and John Davidson and Patrick Leonard, Assistant Directors of the Commission. Two specific agreements were forged during these discussions. FWAG agreed to play a part in setting up "link farms" as a FWAG extension of the demonstration farm idea. Secondly FWAG and the Countryside Commission agreed to collaborate on a seminar for agricultural college principals on conservation and agricultural education, out of which emerged eventually the Conservation in Agricultural Education Guidance Group.

Hookway, though, remained unconvinced that membership of FWAG was a correct move for the Commission. In a memo to Davidson he expressed the view that "direct associations with any voluntary action group are fraught with dangers for an official organisation . . . . Why should we establish such a relationship with FWAG and not a dozen other organisations?" (Countryside Commission FWAG Files). Hookway preferred the idea of "standing one step removed" and so claiming credit for any independent support it gave where appropriate. Davidson did, however, attend a FWAG meeting and subsequently spelt out to Hookway at considerable length its constitution and aims. Hookway was persuaded to allow Countryside Commission representation on FWAG, but no further commitment to the Group. In particular, Hookway was unsympathetic towards FWAG's immediate request for financial aid.

At this time the Commission was anxious to establish its standing amongst other government departments and agencies. Although it was the Government's statutory adviser on countryside conservation and recreation, it lacked the autonomous, grant-aided status of comparable agencies and was without executive functions and powers of any moment. It therefore experienced considerable difficulty in exerting its authority and ensuring that it was consulted when other branches of government were taking important policy decisions affecting its field (Cripps, 1979). In particular, the Commission was trying to establish effective and permanent liaison arrangements with the Ministry of Agriculture, amongst other official rural agencies. The Ministry, for example, had failed to consult the Commission during the preparation of the important White Paper, *Food From Our Own Resources* (1975), in spite of its obvious environmental implications. The Commission was led publicly to complain that: "Over the years we have regularly had cause to regret a lack of regard for conservation and recreation in Ministry policy and practice: we have tended to make more progress dealing with the private farming and landowning organisations directly" (Annual Report 1978-9, p.2). Thus the Commission was loath to see its pressures for adequate liaison with the Ministry deflected through FWAG.

Hookway declared in a memo in February 1978 that "our main policy line is that FWAG should be replaced by some inter-agency liaison committee to cover the official interest". Davidson, too, expressed concern about FWAG becoming "official" if NCC and Countryside Commission funding were granted. The Commission, indeed, actively considered allowing FWAG to fold: clearly the likely alternative if public money were not to be forthcoming. In the event, the NCC tired of waiting for a common response with the Commission and awarded a grant to FWAG in the early summer of 1978, which tided FWAG over its immediate financial difficulties. Some of the Commission's staff felt that it had been embarrassingly upstaged.

The question of Countryside Commission support for FWAG at this time became linked to the outcome of the Strutt Report. The Commission warmly welcomed the report's recommendations seeing them as a "milestone in the evolution of agricultural policy and a remarkable endorsement of many of the ideas we have been advancing as well as much of our experimental work" (Annual Report 1977-78, p.37). The Government set up a working party of officials from MAFF, the DoE and the Welsh Office to consider the report's implications and the Commission was invited to contribute to these discussions. The Commission saw as a major issue the redirection of ADAS and urged that its role should "primarily be a promotional one to make farmers more generally aware of environmental conservation issues and recreation opportunities and to encourage them to look increasingly to local authorities and the specialist agencies for advice" (ibid.).

It was anticipated within the Commission that if these reforms were introduced FWAG would be superseded. Clearly the Commission did not wish to 'back a loser', and Eric Carter, the then FWAG Chairman, was informed in August 1978 that the Commission had postponed any possibility of funding FWAG until the Government's response to Strutt was apparent. Carter delayed nearly two months in replying to this letter, and when he did, he drew attention to the continued liaison between local FWAGs and Countryside Commission officers and the fact that national FWAG was continuing its programme of establishing county groups notwithstanding any uncertainties about the implementation of the Strutt report. He also pointed out that FWAG had not ignored landscape considerations in its work.

The terms of the relationship between FWAG and the Countryside Commission shifted decisively following the change of government in 1979. Expectations of a major follow-up to the Strutt report were frustrated, as were hopes long nourished within the Commission for a greater and more direct role in land-management decisions affecting conservation. The Commission found itself in a much more adverse, if not hostile, political climate – one in which it was necessary to show responsiveness to private and voluntary sector initiatives, rather than bureaucratic indifference, indecision, or pretensions to aggrandisement. In September, Muriel Laverack, an Assistant Director of the Commission, appealed for a speedy decision on FWAG's grant application. Her memo is worth quoting at some length:

> I am unimpressed by our seeming inability to settle this matter over so long a period . . . . The Commission are generally pro-FWAG, particularly since we and it have developed regionally. Such reservations as we have seem to rest on the wildlife orientation, the personal style of the FWAG Adviser, and reluctance to see it become an executive body or a channel for grant aid. The first could be met by pressing for the term 'conservation' to replace 'wildlife' in the title (as it has done already in the orientation of FWAG's work); the second is less important, surely, as FWAG develops on a county basis and may be ignored? The third is within our power to avoid, by practice and/or negotiations . . . . I am personally becoming worried about a slight tendency I see developing against unilateral decisions, against putting our money where our convictions are, regardless of what others do. NCC have emerged better than we have from the past year, publicly thanked at the FWAG AGM for filling the gap left by the Commission. Given our investment in Demonstration Farms and NAL, I see support for FWAG as money well spent. (Countryside Commission, FWAG Files)

Hookway, however, was still keen to explore alternatives, such as funding FWAG purely as a voluntary body if NCC officials would agree to become observers; or reviving the idea of setting up an alternative conservation advisory unit with MAFF and the NCC. But FWAG's identity was by now well established and, as Laverack pointed out, the proposal for a joint advisory unit comprising the official agencies was "contrary to the prevailing mood among Ministers, Departments and agencies for stepping up support for the *voluntary* sector" (Countryside

Commission, FWAG Files). In any case, by the Autumn of 1979, contributions from other sources had brought FWAG some degree of solvency, and it was beginning to turn its fund-raising attention towards its local advisory work as well, and this was where the Countryside Commission's role might be easier. Already the Commission was helping fund the appointment of John Hughes as a farm conservation adviser working for Gloucestershire FWAG (see pages 76-79). In November, Laverack attended a meeting with representatives of national FWAG. She made it plain that, to the Commission, the local FWAG initiatives were of the greatest interest, though a grant of £1000 for 1980-81 was also agreed for the national organization.

The Countryside Commission, more than any other agency, was aware of the potential value and shortcomings of the new Government's single-minded commitment to the voluntary approach. For more than a decade it had been promoting experimental projects and studies in countryside management. Since the mid-1970s the limitations as well as the strengths of countryside management had steadily become apparent. Whilst, on the one hand, it proved particularly effective in tackling tactical problems, such as small-scale landscape improvements, the provision of minor recreational and conservation works, and easing of local frictions between farmers and visitors, on the other it proved incapable of dealing with fundamental conflicts between land users. Nor was it able to hold out in the face of economic pressures or structural changes in the rural economy. Moreover, because most farmers and landowners were unwilling to relinquish their right to improve or develop land for any considerable length of time, management agreements tended to offer only short-term security for the public interest or public investment, and could prove expensive. Indeed, it was becoming evident that countryside management had had little demonstrable impact on agricultural practices or agricultural management (Countryside Commission, 1984).

Within the Commission it was recognized that countryside management was no substitute for reform of the agricultural and planning policies which structured the development of rural areas and thus determined the fate of the rural environment. Countryside management could only be effective if policies favouring the integration of conservation with farming processes were adopted as part of an integrated and comprehensive approach. The Commission made its views known to the Government in its response to the consultation papers that preceded the Wildlife and Countryside Bill (letter from the Director of the Countryside Commission to the Department of the Environment,

9 October 1979; letter from the Chairman of the Countryside Commission to the Secretary of State for the Environment, 2 November 1979). However, its calls for the strengthening of conservation safeguards, back-up powers to management agreements and the reform of the agricultural support system were ignored.

Nevertheless, the Countryside Commission was pressed into giving its full backing to the legislation and the voluntary approach. It had no option. As the Commission ruefully acknowledged, "we exist at the will of government" (Annual Report 1978 - 79, p.1). The Government's line was reinforced by ministerial pressures and other sanctions. Moreover the Commission was implicated in the general vilification of quangos meted out by some of the Government's hardline supporters. For a period its existence seemed in the balance as the Secretary of State for the Environment reviewed the need for all such advisory bodies within his ambit. The NFU and the CLA seized the opportunity to settle some old scores and made various public statements critical of the Commission's role in the countryside. The Commission survived but suffered major budgetary and manpower cuts: its staff complement was reduced by a quarter between 1980 and 1983. The Government also used its patronage powers to discipline this and other agencies. For example, of the new appointments made between 1979 and 1983 to the governing councils and national advisory committees of the NCC and the Countryside Commission, a majority had farming, forestry or landowning interests, and most of them had served or were still serving in an official capacity with the NFU, the CLA or their Scottish or Welsh equivalents.

The seal was set on the new order with the simultaneous retirement of the Commission's Director, Reg Hookway, and the appointment of Derek Barber as its new Chairman in January 1981. Its annual report for that year deplored the polarization of opinion that had occurred during the passage of the Wildlife and Countryside Bill, and pledged the Commission's full support for the voluntary approach and for promoting co-operative action in the countryside. FWAGs were held up as "an outstanding example" of what practical co-operation could achieve. "The aims of the groups closely match our own", claimed the report in announcing that the Commission intended "to extend and develop our links with FWAG, particularly in encouraging their work in offering advice on conservation on farms" (Annual Report 1980-81, pp.3 and 23).

Under Barber's direction plans were laid to shift a major part of the Commission's grantable resources to FWAG. Such a development was in keeping with the continued pressure from the Government to reduce the

Commission's size and executive functions. At the start of 1981, the Secretary of State for the Environment asked the Commission to scrutinize all areas of its activity with a view to reducing wherever possible its direct involvement and staff commitment, and to increase its efforts to work with and through private bodies in the commercial and voluntary sectors. This precipitated a far-reaching revision of the Commission's priorities, involving a shift of emphasis towards conservation and away from recreation and access, planning and development control, and work with schools; and towards advisory and demonstration work and away from research and development. The focus of its advisory role to Ministers, it was decided, should be on how national policy could be adapted to help remove sources of conflict between different countryside interests. The subsequent prospectus (Countryside Commission, 1982) setting out the Commission's policies and priorities took as its theme "conservation through co-operation" and gave particular emphasis to "supporting the voluntary and private sectors – key partners in countryside management". Once again, support for FWAGs was singled out as the centrepiece of an approach predicated on "co-operation with the farming and landowning community rather than confrontation".

In 1981 the Dartington Amenity Research Trust was asked by the Commission to appraise the work of the professional advisory services for farmers offered by the Gloucestershire, Somerset and Suffolk FWAGs, and the subsequent report strongly endorsed the appointment of full-time advisers to county FWAGs (see pages 77-79). Doubts still remained amongst Commission staff, however, concerning the specific objectives and organization of FWAG. Keith Turner, the staff officer most closely involved in negotiations with FWAG, summarized them and put forward a tentative solution:

> During the last 12 years of Jim Hall's reign FWAG has made a useful contribution to creating more awareness about farming and countryside issues – albeit more about farming and *wildlife,* rather than a wider approach embracing landscape, historical aspects, access, etc. The development of county FWAGs has been an obvious, but extremely variable development. Some are active, competent and dependable; others are a dead loss . . . . But the best of these groups are only a talk shop organising occasional open days on suitable farms. They do very little else . . . . Our NAL projects are better . . . . I suggest the National FWAG *reorganises* itself into a Trust-like concern to drum up money . . . . The National FWAG

should be rather like the Woodland Trust and our (and NCC's) grant should be channelled into assisting FWAG to raise money to finance advisers on the ground via *reconstituted* local FWAGs, and be on a sort of "payment by results" basis . . . . The present "constitution" of FWAG is weak, ill thought out and inappropriate to the developing role that is needed. (Countryside Commission internal memo, July 1981)

National FWAG rejected the suggestion of changing its existing structure lest a more formal constitution inhibit the liaison and co-operation which the Group promoted and the top-level informal representation that it was able to attract. But it did accept the Commission's argument that its informal constitution made it difficult for it to receive sums of money at both national and local levels. The decision was therefore taken to establish a FWAG charitable trust to provide a service to potential donors and to county groups (FWAG Minutes, 15 July 1981).

The way was now clear for the Countryside Commission to award a grant with certain conditions attached. These were spelt out, for both the Commission and the NCC in a letter to FWAG Chairman, Wilf Dawson, from the Countryside Commission's Grants Officer in September 1981. The offer of favourable consideration of a formal request for £7500 was made provisional on the grant being paid specifically for FWAG to investigate and report its most appropriate role in the aftermath of the Wildlife and Countryside Bill, how this would relate to the work of other relevant bodies, and ways in which future funding might be achieved to enable national and local FWAGs to become financially self-sufficient within three years. These proposals were readily agreed to. As a consequence a meeting between representatives of FWAG, the NCC and the Countryside Commission took place in February 1982. This meeting marks something of a watershed in that it was agreed that the emphasis of FWAG work needed to change "from one of *discussion* of the issues between farming and conservation to one designed to *secure action and protection* for landscape and wildlife features on as many farms as possible". It was agreed that the furtherance of this aim through encouragement of local groups would become a high priority for the national adviser.

His progress report of November 1982 on the development programme detailed the meetings, conferences and field visits in connection with the 1981 Act which FWAG had organized or assisted with. But, in line with the new priority, its main emphasis was on the future development of County Groups and the steps which had been taken to establish a Charitable Trust to meet the demand for future funding. In March 1983 the Commission offered a grant under its approved policy for Development Grants to Voluntary Bodies to the Farming and Wildlife Trust Ltd. towards the cost of establishing and running it for 3 years and for the support of the Conservation in Agricultural Education Guidance Group (see page 48). It also made a commitment, under its Countryside Adviser Scheme, to offer grant aid for the appointment of advisers in the counties, covering half the costs during the first three years and a progressively diminishing proportion over the following three years. Before an adviser could be appointed, it was agreed, the Farming and Wildlife Trust should require the local group to raise a quarter of the costs of employing the adviser over a period of three years with the Trust itself making up the remaining 25 per cent from national funds. An initial fund raising target of £500 000 over three years was set: sufficient, with Countryside Commission grant aid, to make possible the appointment of up to 30 advisers within the period. John Hughes was confirmed in Gloucestershire under the new scheme and Alison Osborn was appointed in Wiltshire in the Autumn of 1983, soon to be followed early in 1984 by Paul Cobb in Kent. The first FWAG Advisers' meeting was held on 23 January shortly before the formal launch of the Trust on 16 February at the Royal Society of Arts in London.

Graced by the presence of HRH The Prince of Wales who, as Duke of Cornwall and one of the country's leading landowners, had made the first donation to the appeal, the launch was an occasion for affirmations of support, calls for financial backing and worthy sentiments regarding FWAG's role in promoting the voluntary approach. FWAG offered farmers, said National FWAG Chairman Norman Moore sounding a note of caution, "their last chance to regain, in the eyes of the nation, their traditional role of guardians of the countryside." On a day when the six speeches were inevitably precisely timed only Countryside Commission Chairman, Derek Barber, departed from a prepared text. Speaking after NFU President Sir Richard Butler he implicitly recognized the moral solidarity and social control so necessary to the voluntary approach. Two days previously the NFU had issued a press release drawing attention to the criticism and warnings which had been directed at the "mavericks within our midst" at the NFU's annual meeting. At a time when farmer bashing had, he acknowledged, become a popular field sport, Barber seized the opportunity to appeal directly to Butler and other leaders of the industry to "take rather firm steps to deal with those who are brutal, insensitive ... and all who work to destroy the voluntary co-operation ... which we are trying to knit together."

Such moments did little to compromise the overwhelmingly celebratory character of a day which represented a substantial success for the Trust. As a result of personal approaches to a selected list of leading farmers and landowners, agricultural companies and charitable trusts it was possible to announce that £250 000 had already been promised to the Trust over a period of 5 years mainly under deeds of covenant and, as Trust Chairman, Richard MacDonald, put it in conversation, "The list of sponsors reads like a Who's Who of the countryside". In the light of such sums, however; and in view of the target of £500 000 and Trust Director Wilf Dawson's view that £2500 would be an appropriate sum for a national organization to be contributing, Sir Richard Butler's announcement that the NFU would be increasing its contribution to a sum worth £1500 sounded somewhat parsimonious.

The matching contribution from MAFF, though similarly restrained, was of considerable symbolic significance since resources had previously only been given in kind. Indeed, it was rumoured that the decision to contribute had gone up through four levels in MAFF before finally being taken by the Minister himself, though his letter, prefacing the glossy Trust Prospectus and expressing delight that the Ministry had been asked to become one of the founder patrons, gave no hint of such protracted deliberations.

The first year of the Trust's existence saw over £1 million brought into the FWAG movement. In November 1984 its Director was able to report that the gross total raised or promised over a five-year period, including the annual subscription payments to National FWAG, was in excess of £700 000. National grants from the Countryside Commission, Countryside Commission for Scotland and the NCC took the figure to £755 000 and, in addition, the Trust was contracted to receive about £400 000 over the next few years for Farm Conservation Adviser appointments. Fund-raising at the national level had enabled the appeal target to be passed in pleasingly brisk fashion and progress with the appointment of advisers had been similarly rapid so that by January 1987 36 were in post: 32 in England, 3 in Scotland and 1 in Wales (Fig. 4.1).

In view of the prominence accorded to FWAG in Ministerial and other policy statements and its high profile in the post *Wildlife and Countryside Act* period, the buoyant nature of national funding, where the Trust has concentrated its efforts, is not altogether surprising. At the local level, in contrast, the situation is patchy. In Wales, for instance, there are still only 3 advisers and a recent major fund-raising appeal was a failure. The generally apparent reluctance of farmers and local ancillary businesses financially to support FWAG is indicative of a continuing lack of awareness that is one of a number of factors which continue to give the Countryside Commission, in particular, cause for concern.

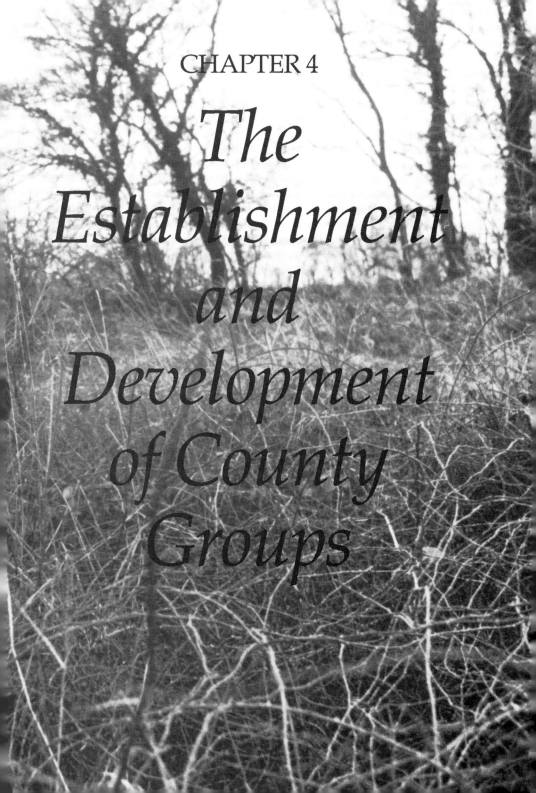

CHAPTER 4

The Establishment and Development of County Groups

Introduction

From the outset FWAG had an evident interest in developing its contacts with the farming community, and in the early 1970s it tended to encourage any local initiative which seemed to conform with its aims. Thus a number of conferences and exercises were supported, as well as various attempts to bring farmers and conservationists together in different parts of the country. In addition, Hall undertook a strenuous round of talks and cultivation of local contacts. The widely circulated FWAG leaflet, 'Farming with Wildlife', set out things farmers could do to help wildlife on their farms, with the emphasis on leaving natural features like ponds, rough corners and banks wherever possible; tree planting in field corners; and avoidance of spraying in such places as hedge bottoms. A FWAG exhibit was prepared and taken to the major agricultural shows and other farming events. These efforts certainly stimulated interest, but in turn they revealed the need for more detailed and specific information and advice if farmers were to be encouraged to make positive changes to their farming practices in response.

At a FWAG meeting in October 1970 the NFU's representative, Michael Darke, stressed the need for a good source of conservation advice for farmers. Though members of the Group were sympathetic, no clear idea emerged on how this might be provided for and what role, if any, FWAG should play. Darke suggested that interested farmers might come together to assist in establishing such a service. But others expressed scepticism about the quality of advice given outside the official channels provided by the NAAS and the Nature Conservancy. The SPNR was at the time exploring how a professional advisory service of the kind required might be set up and linked to the county trusts. Initially therefore FWAG monitored and offered encouragement to existing MAFF and County Trust initiatives.

MAFF Initiatives

In response to the 'Countryside in 1970' Conferences and after the success of the Silsoe exercise, MAFF pursued a number of its own initiatives, very largely through the influence of the then NAAS Director-General and Chief Scientific Adviser, Sir Emrys Jones, who attended Silsoe, and the Labour Minister of Agriculture, Cledwyn Hughes. In January 1970, MAFF organized a national conference on 'Agriculture and the Environment' addressed by the Minister. Officially the advisory staff of MAFF were encouraged to take account of wider countryside interests when advising farmers. Appreciation courses were held for advisers, and the subject of conservation began to play a larger part in the Ministry's general promotional work, for example, at county agricultural shows. Perhaps the most potentially important development was the promotion of a conservation outlook amongst the Ministry's County Agricultural Executive Committees (CAECs).

As a means of implementing agricultural policy CAECs have a long history dating back to before the First World War. However they came to particular prominence when the Ministry of Agriculture took over direct control of them from the county councils at the outbreak of the Second World War. The success of the 'War Ags' in stimulating the drive for production led to their continued use after the War. The 1947 *Agriculture Act* provided for the Minister of Agriculture to appoint five members of his choosing, one being a member of the local county council, and to select the rest from nominees put forward by the three interests involved – three farmers' representatives, two workers' representatives and two landowners' representatives. In practice this meant a selection process operated by the National Farmers' Union, the National Union of Agricultural and Allied Workers and the Transport and General Workers Union, and the Country Landowners' Association.

Through the 1940s and 1950s the committees played a key role pushing for improved standards of farming and landholding, in administering farming subsidies, grants and regulations, and in providing technical advice to farmers. However, by the 1960s, they had lost most of their powers and functions to the National Agricultural Advisory Service (NAAS), which had been established initially to work closely with the CAECs. The committees continued in their role as sounding boards for local farming opinion, also giving occasional assistance in mediating any disputes between local farmers and NAAS officers, but there was a clear need to find new outlets for the committees' work. The setting up of conservation or environment sub-committees provided just such an outlet. MAFF encouraged and promoted this departure to demonstrate its concern for conservation. Initial moves were made by the Minister of Agriculture in 1968-69, and in 1970 he asked all CAECs to set up Environment Groups to keep a watching brief over rural environmental issues.

The response was patchy but a number of Committees moved to establish such groups as well as to initiate conferences and working parties on the environmental concerns and responsibilities of agriculture.

Somerset went so far as to produce a report on 'Agriculture and the Environment'. Hall, as a member of a CAEC District Committee, paid close attention to these developments. He was particularly impressed by the Buckinghamshire and Lincolnshire groups. Buckinghamshire, for example, produced a report clearly dominated by an agricultural perspective in which little mention is made of habitat loss or species protection, (Buckinghamshire Agricultural Executive Committee, Agriculture and Environment Group, Report 1971). It reiterated a view commonly held at the time that hedgerow removal had passed its peak and was no longer a problem. It accepted that trees were "more of a nuisance to agriculture than a benefit", although it recommended corner planting when hedgerow removals took place. It called for greater agricultural representation on water boards; stressed the need for some re-routing of footpaths; and urged less agricultural and non-agricultural littering of farmland.

Introducing a progress report to FWAG members in October 1971, Hall identified the Environment Groups as offering the best potential for providing information to farmers and liaison between farming and conservation interests at the local level:

> The most promising overall approach is to secure the setting up of "Environment Groups" under the aegis of the Ministry of Agriculture through ADAS. It would be within their powers to provide an overall farming/wildlife advisory service as well as keeping a watching brief on the "health" of their countryside.

Two months later he reported:

> Through the Ministry of Agriculture I am trying to encourage the spread of County Environment Groups. These are composite bodies of ADAS, Farmers, Naturalists, etc., who should make themselves fully conversant with each other's problems. They have a watchdog role to anticipate environmental dangers through agriculture primarily but also through other uses of the countryside. They would be an ideal medium for creating advisory services. They would use existing expertise but ensure that advice was well done.

A major obstacle arose from the fact that the new Conservative Government had initiated a considerable reorganization and reduction in personnel of the former NAAS. In 1971 it was absorbed into the new

Agricultural Development and Advisory Service (ADAS), and this involved cutting and streamlining many of its functions. It was thus hardly a propitious time for the Service to be taking on new tasks and, at a meeting in July 1971 with ADAS personnel in the Eastern Region, Hall was unable to convince the officers of the need for County Environment Groups. Inevitably, therefore, in many parts of the country progress in establishing and developing these groups was frustratingly slow. Even within FWAG, not everyone shared Hall's enthusiasm for the County Environment Groups. They were very much the creatures of the agricultural lobby, and their outlook on environmental issues inevitably reflected farmers' views and interests. Derek Barber, in particular, expressed concern at their "top heavy state" and the possible confusion amongst farmer members about their role and responsibilities (FWAG Minutes, 2 November 1971).

He therefore urged the need for independent liaison groups. The issue was settled when in 1971, MAFF announced the abolition of the CAECs. Although their original functions had all but lapsed, it is possible that they could have found a new role in promoting local conservation liaison and advice. FWAG members hoped that something might be salvaged from the various local conservation initiatives that had been launched under MAFF's auspices. Lofthouse of ADAS reported to them that once the CAECs were legally disbanded, ADAS would be free to arrange "Environment Groups" or "Conservation Groups" of its own accord, and many areas were ready to resurrect the idea. It was clear, however, that the disbandment of the committees would leave a vacuum and that the County Environment Groups would need both their own separate organizational framework and a stimulus beyond that which local ADAS officials could provide.

How to Reach Local Farmers?

Despite this recognition of a potential vacuum there was no clear agreement with FWAG over what should fill it. In reaction to suggestions from Barber and Hall that local and independent advisory groups were needed, Norman Moore pointed out that the Nature Conservancy was equipped to answer farmers' queries and was the appropriate body for such work. At a previous meeting Moore had reported that "methods of improving consultation and liaison procedures between the NC, MAFF, NFU and are the voluntary bodies who are concerned with conservation on farmland being explored" (Paper entitled 'The Nature Conservancy and Agriculture', 26 April 1971). To this end, the Agricultural Habitat Liaison Group was set up (see Chapter 3), an Agricultural Liaison Officer (Michael Woodman) had been appointed to the Nature Conservancy's Toxic Chemicals and Wildlife Section and practical leaflets for farmers explaining how and why different habitats should be maintained on farms were being prepared.

Hall, however, was not convinced that this was exactly what was needed. As he has explained: "Nature Conservancy officers had no training in agriculture and were the last people I wanted to visit farmers" (personal communication, 27 November 1986). In a report of December 1971 on "Advisory Services" he argued that "some advisory work is desperately needed to tell farmers about wildlife possibilities on their farms and it has to be given in full recognition of the agricultural and economic considerations, to be acceptable by farmers." Conservation organizations, Hall implied, might play a supportive role but were not suited to be the main channel of advice to farmers. In a subsequent paper, prepared in March 1972, on 'How the FWAG sees the Role of the Naturalists Trusts in the reconciliation of the farming/wildlife conflict', he suggested that the first step Trusts should take was to inform themselves about the situation facing farmers. Armed with a better appreciation of farming they might then explore how to awaken interest in conservation amongst farmers. With many Trusts seeking better contacts with local farmers and liaison with ADAS, Hall's reflections inevitably provoked a strenuous rebuttal from Wilf Dawson of the SPNR, who urged Hall to improve his contacts with Trusts, to learn more of their activities on the ground. Dawson was forced to admit, however, that resources prevented the Trusts from offering a full-scale advisory service for farmers.

At the root of these disagreements were not only different assessments of the nature of the conservation advice that was needed but also contrasting views regarding the legitimate function and scope of various groups and organizations. Hall for his part envisaged a much more salient and direct role for FWAG in promoting conservation advice for farmers, reasoning that "eventually our degree of success will rest in how close we can get to farmers themselves" (Hall, Progress Report, 15 October 1971). He began writing to NFU County Chairmen drawing their attention to the existence of FWAG, and encouraging them "to consider what their President, Henry Plumb, meant when he assured the European Conservation Year (1970) Conference that he 'accepted custodianship of the land on behalf of farmers' " (ibid.). Unfortunately Hall's promptings of NFU branches to get more actively involved in conservation issues and to establish direct lines with the County Naturalists' Trusts seemed to stimulate little response.

There were some promising local developments, however. In 1972, for example, farmers in the East Riding of Yorkshire formed a liaison group and though initially it was quite independent of FWAG, Hall encouraged and took an interest in its development. In March 1973 he reported:

> The Working Party between farmers and naturalists at Bishop Burton in the East Riding of Yorkshire, makes some progress. Some farms are getting individual advice but the biggest single step forward would be to reduce the amount of hedge maintenance in the East Riding. Average size is 30 inches high by 9 inches wide.

The work of this group stimulated FWAG to encourage similar developments elsewhere and, as the Humberside FWAG, it is now recognized as the first of what was to become by the mid-1980s a comprehensive network of county FWAGs.

The idea of local counterparts to FWAG was first mooted at a FWAG meeting in November 1972 by Woodman of the Nature Conservancy. He suggested "a series of local FWAGs who might advise the parent Group and initiate action on its behalf." At this meeting Hall had also reported on the efforts he was assisting of four West Midlands' Naturalist Trusts and the Nature Conservancy to develop co-operative links with ADAS. The meeting had gone on to endorse the potential of the naturalists' trusts to furnish an advisory capacity at a local level. In an attempt to resolve these competing pressures and provide more detailed guidance for Hall, a small steering committee was established, under the chairmanship of Derek Barber. The first meeting took place in January 1973. Agreement was reached upon a plan of campaign, which included identifying people who might promote co-operation locally, and come to act as local FWAG representatives. Considerable emphasis was placed on the contribution of the county trusts, and the possibility of expanding their interest in agriculture so that they might come to provide information and advice to farmers.

One significant development was the offer which saw John Trist, a retired NAAS adviser and a founder member of FWAG, become the first local FWAG representative, in Suffolk. A successful farming and wildlife exercise had been held in the county at Letheringham with the support of national FWAG in 1973. The event was organized by a local committee comprising representatives of ADAS, the CLA, the East Suffolk

Agricultural Institute, the Chadacre Farm Training Centre, the Forestry Commission, the NFU, the NCC, Suffolk County Council Planning Department, the Suffolk Trust for Nature Conservation and the Suffolk Naturalists' Society. Trist had been involved and hoped to follow it up. However the County Council also proposed that the exercise working party should stay in existence and offered to service it. Hall, Trist and members of the Ministry feared a loss of identity and farmer suspicion as a result. So the idea of a working party under FWAG's auspices was conceived as a means of countering the local authority initiative, and it was given the title, the Suffolk Countryside Committee of the Farming and Wildlife Advisory Group. Local exercises elsewhere gave rise to similar liaison groups to discuss conservation problems. Typically they included representatives of the Ministry of Agriculture, the NFU, the CLA, the NCC, the local county trust, the county planning authority and the local agricultural college.

In December 1974 Hall prepared a short paper entitled 'The need to operate working parties constituted on the pattern of the national body, designed to carry out the FWAG concept at local level'. Hall was anxious to nurture the various groups growing up within FWAG's ambit but also to relieve the burden of servicing them:

> It is imperative that ways be found of making these working parties less reliant on the parent body and to do so they must have their own secretaries with funds to meet administration and travelling expenses as a minimum. It is visualised, for example, by the Eastern Region of the Ministry of Agriculture, Fisheries and Food, that there should be a FWAG-inspired Working Party in each of the counties in its region, and that they should each have their own secretaries responsible to FWAG, so removing any risk of rejection by appearing to be supported in particular by any one of the organisations concerned . . . . The secretary will in effect be a local FWAG adviser as he becomes conversant with his subject.

Hall was also keen that FWAG should keep a hold of these local groups. The problem was how to give coherence and direction to what were sporadic local iniatives. Reporting in 1974 the imminent establishment of two groups in Yorkshire, following in the footsteps of the already existing North Humberside group, Hall remarked: "we must keep close control in some way or other, else they will switch to the more easily understood facts of the countryside such as landscape and recreation".

Promoting Local Groups

It would be a mistake to suppose that the formation of local groups had come to dominate the work of FWAG at this stage. On the contrary, Hall's work continued to reflect a wide range of initiatives which included farm exercises, developing links with agricultural colleges, publicity, and so forth. Even so by November 1975 Hall was able to report progress with the formation of local FWAG-type committees as:

(1) In existence: North Humberside, North Yorkshire, West Yorkshire, Cambridgeshire, Gloucestershire, Surrey, Suffolk, Bedfordshire, Northamptonshire.
(2) Every chance of being formed: Essex, Sussex, Wiltshire, Lancashire, Leicestershire, Norfolk, Herefordshire.
(3) Possible: Nottinghamshire.

No formal decision had yet been taken by national FWAG on the appropriate constitution of such groups, nor on the need to embark on an active campaign to encourage their formation. An important step in coordinating these disparate efforts was the instigation of an annual meeting of county committee secretaries. The first, which took place in April 1976, also gave Hall the opportunity to invite interested individuals known to him in counties where no FWAG group had yet been set up. In the event representatives from the eight counties with groups were joined by individuals from a further seven counties.

That national coverage was not envisaged at this time is reflected in the agreement reached with the Countryside Commission, that the Commission's Demonstration Farms would not be located in areas covered by FWAG groups and vice versa (see page 51). Neither side fully adhered to the agreement, which in any case was rescinded in 1978. The Commission proceeded with its plans in Essex and placated FWAG concern by inviting Hall to join the local steering group; Essex FWAG gave the project its blessing and provided the nucleus for the steering group. In Durham a joint FWAG-Demonstration Farm group was organized. In Oxford and West Norfolk the Commission proceeded alone, but in Somerset, FWAG set out to form a local group notwithstanding the possibility of a demonstration farm project.

The Chairman's Working Party, formed in April 1976 to review FWAG's role and its future (see pages 31-2), suggested that the creation of new committees should become high priority. As a result, FWAG's terms of reference were amended at its meeting in November 1976 to include the aim of establishing a "system for liaison and contact between farming and conservation interests at local level with a view to providing practical advice on wildlife conservation to farmers". The meeting agreed that the formation of local groups and the provision of more support for them should be top priorities for FWAG, and it went on to suggest what form this support might take: "an annual meeting of local secretaries with representatives of the National Committee; the copying of the National Committee minutes to local secretaries; the encouragement to local committee chairmen and secretaries to feel that they can turn for help and support to the National Committee".

With the new emphasis on local groups, more deliberate steps were taken to stimulate their formation. These were not always successful and indeed there were a few false starts. Jim Hall reported in April 1979 that "Shropshire had recently set up a FWAG committee/branch albeit reluctantly" (FWAG Minutes, 5 April 1979). Some groups lapsed, including those for Derbyshire and Bedfordshire. In Devon, local NFU opposition frustrated the start of a group for five years after one was first proposed in 1976. The initiative was only revived when a number of farming and landowning members of the Committee of the Exmoor Society resigned because of the Society's critical stance on moorland reclamation in the national park, and the idea was mooted of a North Devon FWAG to provide a platform more sympathetic to farming interests. A recurrent element in the wariness of local farming opinion towards the creation of a county FWAG was that any such group might be hijacked by conservationists and provide a platform for the critics of agriculture. If anything, though, the real risk was the opposite. In many instances, NCC, RSPB and county trust personnel found it difficult to allocate time to FWAG, whereas in Dorset the Secretary of the County Naturalists' Trust turned down an invitation to instigate a group, reporting that she was satisfied with existing liaison with the NFU County Branch. At the end of 1976 Hall analysed the state of the county groups in the following terms:

> Their success is by no means certain; the conservation input is the weakest link in almost every case and has to be strengthened. This must be a priority for the conservation interests on the Group.

But local suspicions had to be overcome. The potential difficulties were apparent from the beginning in the North Humberside group. With a strong base in both the local farming community and the Bishop Burton College of Agriculture conservationists felt the group's position to be weak and compromised. In turn the group's first secretary criticized the conservation organizations for their lack of involvement. North and West Yorkshire also reported the reluctance of naturalists to be involved in the early days. Surrey was somewhat exceptional in that the early running was made by members of the RSPB. In other counties the establishment of a committee of sufficient size and enthusiasm often proved difficult – only three people attended Leicestershire's third meeting in 1976 for example. Essex and Gloucestershire were amongst the groups that achieved wide support from the start.

Before a group was started there had to be a meeting of interested parties and for this to be a success, Hall usually had to lobby vigorously beforehand to get the right people and the right support. The key individuals in getting a county group off the ground were usually the chairman and the secretary. The pattern was quickly established that the chairman should be a well established farmer sympathetic to conservation and prominent in local farming or landowning circles. Hall usually looked for someone who was a leading figure in the county NFU so that news of what was afoot would reach NFU members. In many cases, there would be an obvious candidate – such as an active farmer-member of the council of the county naturalists' trust, or someone who had participated in a FWAG exercise. County secretaries were more diverse. Often when a group was getting off the ground, the principal of the local agricultural college was contacted to see if a member of his staff would be willing to act in this capacity. In other cases secretarial support was provided by ADAS, the county trusts, county planning departments and rural community councils.

Inevitably there was some diversity in the interests and activities of local groups. North Humberside, reporting in 1977 on its first five years of activity, highlighted the preparation of three farm conservation plans, one for an ADAS experimental husbandry farm and another for an LEA College Farm. A number of farm visits had been arranged largely for the benefit and experience of the group members themselves, and a local exercise attracted one hundred people, of whom only twelve were farmers.

The Group had also produced its own advisory booklet on tree planting and had adopted a relatively high public profile commenting on issues in the locality, including a number of non-agricultural environmental issues, such as industrial development, road widening and the construction of the Humber Bridge. Other early groups showed a catholicity of interests although not as broad as the North Humberside group. The Northumberland group, formed initially as the Northumberland Countryside Liaison Group by MAFF in the aftermath of the Cowbyers conference, took on the problems of urban-fringe farming, particularly rubbish dumping. In contrast the Surrey group, which also commenced in 1975, directed its attention at farm wildlife issues more central to FWAG's remit.

As efforts continued to expand the network of county groups and to weld them into a co-ordinated structure for encouraging farmers' interest in nature conservation, there inevitably arose certain national/local tensions. National FWAG regarded the local groups not so much as semi-independent county bodies but as local committees operating under its auspices. It also had a narrower and more instrumental view of their role, certainly than was held by some of the older county groups. National FWAG had to find a balance between promoting its own objectives and maintaining the allegiance of the local groups. Thus it is no surprise that the model constitution prepared for the FWAG secretaries meeting in May 1977 was not immediately acceptable to all groups and Hall was at pains to stress that the proposals had been put forward only to elicit their comments and discussion.

It was at this meeting that the delineation of local FWAG's concerns first became a significant issue. In particular the rather wide-ranging interests of the North Humberside group came under scrutiny. Whereas Jim Hall had supported North Humberside in its action in submitting comments on the county structure plan, the national Chairman (Ian Prestt of the RSPB) felt that "there was a grave danger in comment of this sort, in taking the mantle of the constituent bodies in a way to which they might take exception and which would lead either nationally or locally to their withdrawal from FWAG" (Minutes Local Committees' Secretaries Meeting, 10 May 1977). The Chairman of the North Humberside group responded by appealing for the links with local groups not to become too formal, for his members were "very independent people and would be liable to reject anything that smacked of an instruction" (ibid.).

Subsequently FWAG's management committee in considering the draft local constitution specified the following narrow terms of reference:

> To establish a forum for informal liaison and contact between farming and conservation interests at local level with a view to providing practical advice on wildlife conservation to farmers. (Consultative Paper, prepared by Hall, June 1977)

National FWAG, however, has sensibly not tried to dictate to local groups but has sought influence through leadership, information and informal contact. To the annual meeting of FWAG secretaries was added a regular newsletter begun by Hall in the autumn of 1977. This softly-softly approach has prevented any defections from FWAG though at the price of a certain lack of direction. It should also be added that Hall, who mediated the links between national FWAG and the local groups, took a much more permissive stance than did his colleagues and quietly encouraged the groups to take an active role in local environmental politics.

Hall regularly attended events organized by county groups across the country and received minutes of their committee meetings. He kept a watchful eye on their progress, and provided a steady flow of ideas, information and advice to encourage them to develop their range of activities and usefulness. Innovative projects and events mounted by one group were reported to the others; such as staging a farm competition; provision of articles for local farming programmes; preparation of a lecture/slide presentation for farming audiences; rehabilitation of an unproductive woodland as a demonstration area; persuading an agricultural show society to adopt a conservation scheme; production of a map showing nature reserves and SSSIs for the guidance of aerial spraying contractors; preparation of display units for meetings and shows; production of a leaflet giving practical advice on local conservation problems; and attracting commercial sponsorship. Hall laced all this with his own suggestions and appeals, ranging from topics for conferences and exercises, to tips on how to avoid FWAG displays being relegated to the periphery of agricultural showgrounds.

Developing the Advisory Capacity of Local Groups

The major function of the county groups, promoted by national FWAG, was to stimulate amongst farmers an awareness of wildlife and an interest in its conservation. A number of counties, for example Norfolk and Lancashire, adopted the local exercise as a good way to instigate some sort of educative and advisory work. Others organized farm walks, and some, such as Sussex, were confident enough to publicize themselves as a source of advice. By 1977 the more energetic counties were beginning to think how they could improve their advisory work and take it beyond the small circles of involved and committed individuals. The following year, Jim Hall reported in his newsletter:

> One or two committees are running into trouble because they have generated more interest and more requests for advisory visits than

their secretaries can handle. The answer to this dilemma must in the first place lie with giving the secretary extra help from within the committee, although in the end success may mean . . . a search for funds to employ somebody. (FWAG County Committee Newsletter, Autumn 1978)

The Somerset Trust for Nature Conservation showed the lead in recruiting a farm conservation adviser paid through grants from the NCC and the Ernest Cook Trust. The officer worked to a steering group with a wider membership than the Trust itself, including representatives of ADAS, the CLA, the NFU and the county planning department, and it subsequently became the Somerset FWAG. The person appointed, Miss Dorothea Nelson, had qualifications in dairying and many years experience of farm work, and had a keen interest in wildlife. Her remit was "to promote the conservation of existing features of wildlife interest and value on a selected series of Somerset farms, and encourage and advise on the creation of new wildlife habitats". It was hoped that her work would also reveal details of farm habitats, trends in habitat loss and the opportunities for and obstacle to integrating conservation with farming.

Other county FWAGs quickly found that, as demand rose for their services, the groups' members became overstretched and they were faced with the choice of delegating in some way or pulling back into a 'talking shop'. In 1978, in North Humberside, a Manpower Services Commission's Special Measures Programme gave short-term funding for an NFU/FWAG Conservation Officer and three assistants to carry out practical landscaping and conservation work on farms, such as tree planting, with the intention of establishing a number of demonstration farms around the county, but the scheme folded prematurely when the conservation officer left for more secure employment. Early in 1978 Gloucestershire FWAG also began to seek funds to employ a farm conservation adviser. As with the Somerset appointment, both national FWAG and SPNR were involved in the negotiations to secure funding. The national FWAG Minutes recorded, "ideally the person appointed would be mature and have an agricultural background" (29 June 1978). Funds were obtained from the NCC, the Countryside Commission and the Ernest Cook Trust, and in May 1979 Mr John Hughes, an experienced farm manager and member of the Gloucestershire Trust for Nature Conservation and the Gloucestershire Naturalists' Society, was appointed "to provide advice over the farm gate from a well-advised and experienced farmer, on demand". He was formally employed by the Gloucestershire Trust, which along with ADAS, provided him with a base and office support.

The Suffolk group adopted a somewhat different approach. Rather than seek external funds, the group accepted the offer of two of its members to act as independent consultants. So in 1978 it was able to launch a self-financing service for farmers who required "practical advice of a general nature on procedures to improve the conservation of all aspects of wildlife and the landscape without prejudicing the high levels of farming efficiency". Of the two consultants, Brigadier Fishbourne was a landowner and former CLA regional secretary, and Mr Gordon Clarke was a retired ADAS officer and the Conservation Officer of the Suffolk Trust for Nature Conservation. Their services were offered to local farmers initially at £5 per hour, plus travelling expenses.

To a considerable extent, the Somerset, Gloucestershire and Suffolk schemes established the pattern for the future development of FWAG advisory services. So it is worth examining in some detail their mode of operation, drawing upon an unpublished assessment of them prepared by the Dartington Amenity Research Trust for the Countryside Commission (DART, 1981). By the spring of 1981 after they had been in post for two and three years respectively, John Hughes had visited 180 farms and Dorothea Nelson 154, whereas the Suffolk consultants after two and a half years had covered 55 farms. In each case the practice was to respond to invitations from farmers to give advice rather than for the adviser to initiate contact. The service on offer, though, was well publicized through the local newsletters of the CLA, the NFU, ADAS and other interested organizations; and the advisers spent much time stimulating interest and establishing contacts through farmers groups, Young Farmers' Clubs and Women's Institutes, giving talks at Colleges of Agriculture and presenting exhibits at agricultural shows.

When visiting a farm, the advisers would typically go the rounds with the farmer making a brief appraisal of its habitats, establishing the farmer's circumstances and interest in wildlife, and responding to specific queries. The form and style of the advice given differed somewhat amongst the advisers, reflecting their different remits and the demands made upon them. John Hughes in Gloucestershire tended to offer his advice during the farm visit, relating the opportunities for wildlife conservation to the general running of the farm and its development. Dorothea Nelson in Somerset, in contrast, would usually make a return visit to conduct a more detailed appraisal of the farm's most promising areas in order to prepare guidelines on how these might be preserved and their value as wildlife habitats enhanced. The Suffolk consultants, for their part, would often draw up a farm-wide strategy involving not only wildlife conservation measures but also landscape improvements requested by farmers. A common feature of the role of all the advisers was to put farmers in touch with the agencies that could give more specialized advice and possibly practical assistance or grant aid, such as MAFF, the Forestry Commission, the Nature Conservancy Council, the County Council, the Regional Water Authority, the Game Conservancy and the county trust for nature conservation. At the same time, advisers were, and presented themselves as, independent of any statutory body. The DART report identified this as a key factor in their being able quickly to establish a rapport with farmers. The advisers were "seen as being on the side of the farmers and as being able to evaluate the conservation position within the farmers' frame of reference" (DART 1981, para 106). Crucial to this also was their agricultural background which had obviously been a decisive consideration in their appointment, with much less emphasis placed on wildlife expertise. Indeed none of the advisers had any formal qualifications in natural history or conservation, although some on-the-job training had been arranged for each of them.

However there is no doubting their impact on the farms they had visited. Between 64 and 80 per cent of the farmers had taken positive action following advice; others were contemplating such action or had refrained from deleterious steps. Only between 4 and 11 per cent had shown indifference or hostility. It is likely that the proposals made by the FWAG advisers in these three counties during the first 2 to 3 years resulted in expenditure in excess of half a million pounds on the farms they had visited. The DART report concluded:

> the advisers are seen very much as the 'cutting edge' of the county group concerned . . . . They are achieving results which would not have been achieved by their parent county groups and . . . we believe that they are cost-effective in that they lead to a total shift in resources many times the cost of the appointments; and they achieve permanent changes in farmers' values and awareness – to a much greater extent than the same amount spent, say, on leaflets or contributions to the media. (DART 1981, paras. 86 and 114)

It should be said, however, that these gains were not achieved without overcoming certain difficulties largely inherent in a voluntary organization mounting a professional service. The DART report referred to a "feeling of redundancy amongst those volunteers who formerly gave advice to farmers on an *ad hoc* basis" (DART 1981, para. 88). This was not the only source of internal strain; the other was conflicts of interest when a FWAG member organization objected to what it perceived as too close an identification with a client by the adviser. In some cases, the latter was privy to information, such as sightings of rare species, which the NCC and the county trust would have liked to record, but which was denied to them, even though they were helping to sponsor the scheme, as the adviser felt this would be a breach of the farmer's confidentiality. In other cases, conflict arose when the adviser compromised with a farmer on action which an individual member organization opposed on principle, including, in one example, the possibly illegal felling of trees. In all these cases, the advisers received the ultimate backing of their steering groups.

Apart from such internal strains there was also the administrative and financial burden of operating the schemes. In Gloucestershire and Somerset the burden fell on the county trusts. Perhaps the most serious difficulty, apart from raising the funds in the first place, was the need to juggle cash flows to accommodate the cost of the adviser as most of the grants from statutory bodies were paid in arrears and did not cover all the costs. Continuity of funding was also a problem, with most grant-giving bodies committing themselves only on a yearly basis and seeing their support as short-term pump priming. One consequence was that the advisers' positions were not secure and not well paid. Some of these difficulties did not arise with Suffolk's self-financing consultancy, though Jim Hall did identify a potential danger "in denying a 'free' service to those who would regard the 'following' of advice as their payment and the giving of it free as proof that someone else also thought conservation important" (FWAG County Committee Newsletter, Autumn 1978). The success of the Suffolk scheme, however, clearly demonstrated the existence of a suitable number of prosperous farmers already attuned to the need for some conservation advice and willing to pay for it. Most other counties, though, were unlikely to be so favourably endowed.

The development of the three schemes was keenly monitored and encouraged by national FWAG. However, only a small number of county groups were in a position to employ or sponsor their own professional staff. It is significant that by 1981 no other county group had yet followed the lead of Somerset, Gloucestershire or Suffolk (Figure 4.1). Even in those groups which perceived the need for greater outreach to the farming community, opinion was divided over what form this should take, as the example of Essex FWAG shows.

Formed in 1976, the Essex group was one of the most active. It met 5 to 6 times a year, and held a similar number of events such as farm walks, conferences and social gatherings. Its members provided monthly articles for the local NFU journal *Essex Farmer*, and it organized an annual conservation competition. The group discussed a wide range of topics and showed itself to be an important forum in the conservation scene in Essex. In spite of this activity, however, its contact with more than a small number of farmers remained slight. The 1980 conservation competition, for example, attracted only four entries, and the group became concerned that it was doing little in the way of stimulating direct requests for advice from farmers. One method of reaching farmers was developed through the associate membership scheme which allowed any interested person to support the group for a small annual subscription. This approach was not recommended by national FWAG with its anxiety to retain the appropriate balance of interests in local groups, and in any case only 20 farmers joined in the first year.

Preliminary discussions were held at the February 1980 meeting on the possibility of appointing a conservation advisory officer. In October of that year the group set up a working party to consider the possibility of a FWAG consultant being appointed to provide on-farm advice. At the December 1980 meeting the services of two consultants were discussed. Some disquiet was expressed at giving two individuals the group's unqualified approval. Nevertheless it was agreed that their services should be offered to farmers, but that all requests should be handled by the FWAG secretary and some forwarded to other advisory agencies, such as the county council and the NCC, rather than to the consultants.

However the plan ran into considerable difficulties at the next meeting. Attention was drawn to the system operating in Oxfordshire where any farmer seeking conservation advice was sent a list of all sources of advice in the county, including members of the county council, individual FWAG members and consultants. The problem in Essex was that one of the consultants had been asked, in an individual capacity, to join FWAG. Some members felt that this was unfair when there were other consultants operating locally who were qualified to give conservation advice. It was argued further that all members of FWAG, other than farmers, should be representative of one of the constituent bodies. The outcome was that the invitation to FWAG membership was withdrawn from the newly appointed independent consultant, who was present at the meeting. This embarrassing incident, although in no way a reflection of disagreement on the need for advice, effectively postponed further serious consideration of the development of professional advice for some time.

At the next meeting (April 1981) it was agreed that Essex FWAG should continue as "an enthusiastic amateur group" which was "capable of giving advice on a friendly basis". The issue was not discussed again until June 1983. In the meantime the group continued with its usual round of open days, annual conference, informal meetings with special speakers and conservation competition. Although the extent of the group's activities and the depth of involvement of a dozen or so key individuals made Essex FWAG one of the success stories of FWAG's formative years it had to some extent become rather limited in its aims by 1982-3. It had run the same events for a number of years and by its own admission was frequently preaching only to the converted. For this and most other county FWAGs it was evident that it would require external stimulus for them to develop an extensive advisory service.

Expanding the FWAG Network

While some of the established county groups sought to extend their role and utility, the FWAG network continued to expand under strong encouragement from the national group (see Fig. 4.1). Changes in the political climate helped force the pace of development – most notably the politicization of the countryside precipitated by the *Wildlife and Countryside Act* 1981. In 1981, Eric Carter succeeded Jim Hall as the national FWAG adviser and in his first newsletter to the county groups he observed that the debate over the Act's passage had revealed not only "the depths of public concern about the countryside" but also "a lack of knowledge about farming and countryside matters, misinformation and sheer prejudice". He concluded:

> If FWAG did not exist there is little doubt that it would have to be invented following the passage of the Wildlife and Countryside Act. (FWAG Newsletter, Autumn/Winter 1981/82)

Because of its national scope, the effect of the legislation was to raise the profile of farming and wildlife problems everywhere, and not just in those areas where the problems had long been recognized. This gave an impetus to establish county groups where none had existed. The legislation was first announced in 1979, and more than half of the FWAG groups have been formed since then. A number of these 'late-comers' are counties of striking conservation importance, including Cumbria (March 1982), Devon (February 1982), Dorset (May 1981), Gwynedd (March 1982) and

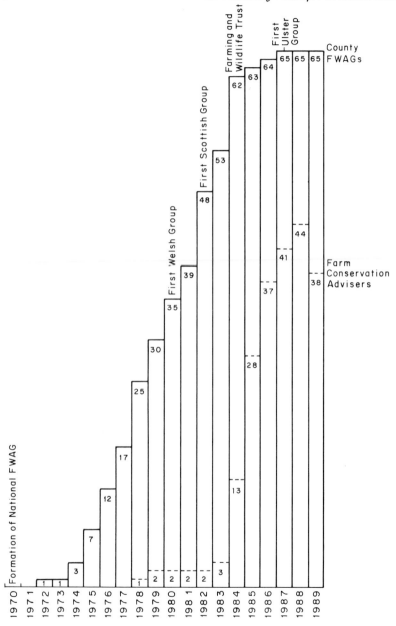

Figure 4.1 Growth of County FWAGs and Farm Conservation Advisers (---). Note: the statistics for County FWAGs refer to the end of the stated calendar year; and for advisers, to the numbers in post at the end of the following year (i.e. the following March).

Herefordshire (November 1981). Even in some of the counties where it was long established FWAG had a low profile prior to the Act.

FWAG now claims complete coverage of England and Wales (as well as Scotland also). This is not strictly true as Radnorshire, in Wales remains without a group. The geographical pattern of growth is of interest. The first groups to be established were all in the eastern counties. These are the counties that have borne the brunt of the post-war revolution in arable production, where the most sweeping changes have occurred to the rural environment through the mechanization and chemicalization of farming. Inevitably, the pressures and the conflicts over conservation have been most acute there; and it is no coincidence that these counties also have the oldest and largest county trusts for nature conservation. Typically they also have high proportions of large, prosperous, innovative and business-like farmers susceptible to FWAG's appeal and with the time, awareness and resources to be ready to take part.

Of course, pressures for agricultural intensification spread well beyond the traditional arable counties during the 1970s especially in western England, the west Midlands and parts of Wales. Then in upland Wales and the north of England there was the added pressure of afforestation. During the late 1970s and early 1980s FWAGs spread to these other regions, with counties with a predominantly big-farm structure generally taking the lead, and some of the counties dominated by small farmers being among the most recalcitrant. Progress within individual counties has often presented a different and more complex pattern. Thus, whereas counties such as Hampshire and Norfolk have found the larger farms and estates most receptive to FWAG, often related to a tradition of landscape conservation on these holdings, other counties, including Kent and Staffordshire, have found most receptive the long-established, smaller farmers, keen to preserve a pleasant working and residential environment for themselves and their families (Reynolds, 1984). In the following chapters we seek to explore how the range of county groups operate.

CHAPTER 5
The Workings of County FWAGs

Introduction

The remainder of this study examines the local operation of the FWAG network. The following two chapters present case studies of Wales and Wiltshire respectively. This chapter provides a contextual overview derived from a postal questionnaire survey of county groups in England and Wales. It was conducted during Summer 1986, and 38 out of 48 groups responded, giving a 79 per cent response rate. The questionnaires were addressed to the FWAG county secretaries. The returns present a snapshot of the structure, personnel and activities of the county groups. Inevitably, they reveal much variation across the country. The timing of the survey coincided with a particular phase in the development of most groups. Whereas a few had employed full-time advisers for more than two years, most had only recently taken this step and some were still contemplating it (Figure 4.1). The survey returns, therefore, present an opportunity specifically to evaluate this important transition and its consequences. The chapter also makes use of the findings of a recent review of the work of FWAG advisers conducted by the Centre for Rural Studies at the Royal Agricultural College (CRS, 1990). Two of us, Philip Lowe and Michael Winter, were members of the CRS team.

Composition and Organization of County FWAGs

Figure 5.1 lists the organizations that are represented on the County Committees. Not surprisingly, certain organizations predominate, and these are mainly the ones represented on national FWAG. However, two of the latter – RICS and BTO – evidently do not have the same degree of *local* commitment to FWAG as do the others. Respectively they were formally represented on only 11 and 3 of the county groups in our survey (several more, it should be noted, had land agents on their committees, and they most likely were members of RICS). Conversely, a number of organizations not represented on national FWAG are extensively represented at the local level, including County Councils, Water Authorities, Agricultural Colleges, Young Farmers' Clubs and branches of the CPRE and CPRW. All of these organizations have considerable interest and some influence in the way the countryside is used and managed, and it seems appropriate that they should be represented on county FWAGs. The absence of some of them from national FWAG seems an anomaly though the CPRE was excluded at the insistence of the NFU and the CLA.

From Fig. 5.1 one can identify a set of core organizations each represented on at least 6 out of every 7 County FWAGs. These are ADAS,

Figure 5.1   Organizational Representation on County FWAGs

Organization	Number of county FWAGs on which organization is represented (out of 38)
ADAS/MAFF*	37
NFU*	37
County Trust*	35
CLA*	34
Countryside Commission*	33
NCC*	33
County Council	32
Forestry Commission*	31
RSPB*	27
Water Authority	24
Agricultural College	24
Young Farmers' Club	21
CPRE/CPRW	16
RICS*	11
Rural Community Council	11
BTCV	9
Ornithological Society	6
Farmers' Union of Wales	5
Agricultural Society	5
Game Conservancy	5
Women's Institute	5
Archaeological Society	4
National Trust	4
Timber Growers Org.	4
BTO*	3
NUAAW/TGWU	3
Agricultural Training Bd.	3
National Park Authority	3
Groundwork Foundation	2
University	2
Royal Forestry Society	2
Others†	1

*Members of National FWAG

†Organizations represented on only one county FWAG included:
Friends of the Earth, Economic Forestry Group, Botanical Society, British Association for Shooting and Conservation, Golfers Association, Institute of Terrestial Ecology, Campaign for Keeping Wales Tidy, British Field Sports Society, Ramblers' Association, Women's Farming Union, Shropshire Chamber of Agriculture, East Yorkshire Conservancy Council, Warwickshire County Museum, Shropshire and West Midlands Show Committee, South Holderness Countryside Society, Hoechst UK Ltd., Murphy Chemical Co.

the NFU, the county trust, the Countryside Commission, the NCC and the CLA. Often some of these organizations are severally represented, particularly ADAS and the NFU. In addition, the NFU, the CLA and the county trusts usually enjoy indirect representation, often on a multiple basis, through their members serving on a county FWAG in another capacity. One group, for example, included as many as 12 members of the county trust.

However 29 per cent (11) of the county FWAGs lacked one of the core organizations, including 16 per cent (6) that lacked two or more. In some cases it may be that informal links, or indirect representation suffice, because many of those involved in county FWAGs are wearing several hats. In a few cases, though, it is difficult to avoid the conclusion that the gaps in formal representation are real deficiencies arising from a failure as yet to involve key agents in the local farming and conservation scene. Half of these incomplete groups are in Wales where FWAG has met with markedly less enthusiasm than most parts of England, for reasons that are more fully explored in Chapter 6. One Welsh county secretary replied, "A response to your questionnaire is not feasible as this group is in a state of suspended animation . . . . A lack of commitment to the concept of conservation within the local farming community may be the prime reason for slow progress". Significantly also, most of the county groups lacking one or more of the core FWAG constituents are amongst the more recently formed, and presumably they are still in the process of establishing themselves.

If the representation of a county FWAG is too narrow, the group may be denying itself vital expertise, crucial local knowledge or influential contacts. Moreover, failure to win a range of organizational support can lead to other weaknesses. For example, 6 out of 11 of the county groups without the full complement of core organizations were without an adviser at the time of the survey and they accounted for half of those lacking an adviser. Representation of two organizations in particular, the Countryside Commission and the county council, seemed vital in generating sufficient resources to appoint an adviser. Indeed, only one group on which the Commission was unrepresented, and only two on which the county council was unrepresented, had a FWAG adviser. Although, with the support and encouragement of the Farming and Wildlife Trust, more groups have since acquired advisers, it may be that some of the organizationally weak FWAGs may be unable to sustain the necessary financial commitment in the medium term.

Certain other variations in representation invite comment. For example, some 11 out of the 38 county groups lacked an RSPB representative, but this omission was compensated for in Buckinghamshire and Gwent through the inclusion of ornithological societies. Nine county groups remained without any representation or ornithological interests including counties such as North Humberside, Dorset and Gwynedd of outstanding ornithological importance.

Representation of forestry interests and expertise also shows great variation. Some of the groups in upland counties with extensive commercial afforestation, particularly the Welsh ones, include a number of forestry organizations. For example, the Montgomery group has representatives of the Forestry Commission, the Timber Growers Association and the Economic Forestry Group. The Welsh groups (like the Scottish ones) have adopted the title Farming, Forestry and Wildlife Advisory Groups (FFWAGs). One English group, Northumberland, has followed suit. In contrast, a few county groups, mainly in central southern England, are without any forestry representation. Though these are areas lacking in extensive commercial plantations the recent shift in favour of encouraging the amenity and nature conservation aspects of woodlands as well as interest in farm-forestry suggest the need for relevant silvicultural expertise even in these counties, especially as advice on what to plant and how to care for trees forms a large part of the work of county groups (see below) and only a few advisers have forestry experience or training. Recent survey work, moreover, has revealed a serious decline in the state of small woodlands on farms through neglect and mismanagement, with only 1 per cent of farmers interviewed having sufficient knowledge to manage their woodlands effectively (DART, 1983).

Another potential gap relates to the representation of water authorities. In many parts of the country their drainage and water management activities have considerable implications for both farming and nature conservation. Furthermore, the growing incidence of agricultural pollution of water courses is a matter of considerable public concern for which the water authorities have ultimate responsibility. Although two-thirds of county groups include water authority representatives, there are some surprising omissions, including some counties where the problems and pressures are particularly acute, such as Norfolk and Lincolnshire.

One type of interest and expertise surprisingly thinly represented was that of game conservation. Only 5 groups included a representative of the Game Conservancy; and one other group, a representative of the British Association for Shooting and Conservation (BASC). Although other county groups undoubtedly include members who are knowledgeable about game, this interest could usefully be more widely represented, and since the survey was conducted it has increased. The Game Conservancy and BASC have accumulated considerable expertise, and for many farmers and landowners an interest in game, whether in pursuit of a personal hobby or an income from the letting of shooting rights, is their point of departure in encouraging wildlife on their land. Both these organizations offer an advisory service to farmers and landowners, and effective co-ordination with the work of county FWAGs would seem desirable.

Of course, which organizations are included in a county group is a matter entirely of local discretion. While this may produce certain anomalies and apparent gaps, it also means that county groups are able to reflect local issues and interests, make appropriate alliances, and involve county loyalties. In consequence, each group has a distinctive organizational make-up. The Lancashire and Cheshire groups, for example, have representatives of the Groundwork Foundation, which is involved in extensive landscape reclamation and improvement projects in the two counties. Most of the Welsh groups include members of the Farmers' Union of Wales, as well as its rival, the NFU. The Gwynedd, North Yorkshire and Cumbria groups include representatives of the respective national park authorities. In Norfolk, Suffolk, the Isle of Wight and Cornwall, the local archaeological society or unit is represented in the county group. This local pluralism extends to the point where certain county groups include organizations whose parent bodies have been critical of the consensual approach pursued by national FWAG (such as Friends of the Earth in the Herefordshire group, and the Ramblers' Association in the Essex group); and others which have been excluded from national FWAG (such as the CPRE in various county groups, and the British Field Sports Society in the Essex group).

Getting sufficiently strong and broad organizational support is important for a county FWAG in establishing its legitimacy and authority and in opening up access to certain organizational resources, including finance and expertise. Equally important is getting an appropriate mix of committed individuals: people sympathetic to both farming and conservation, with relevant knowledge and contacts, able to work together and incite interest amongst local farmers and landowners. FWAG, indeed, stresses that committee members serve in an individual capacity even when chosen in part for their organizational links.

Most committees include several independent members, mostly farmers and landowners, comprising up to half the total complement.

Significantly also, some of the non-agricultural organizations often choose representatives with a farming background or connections. RSNC, in a survey of county trusts conducted in 1984, found that agriculture, forestry and related professions accounted for almost half of the county trusts' non-staff nominees to FWAG committees (Heaton, 1974). Such practices, as well as the number of independent farmer members and the multiple representation of the NFU, the CLA and ADAS ensure that on most county FWAGs agriculturalists and agricultural interests predominate.

With county committees ranging in size from 10 to over 50 with a median figure of 28, the running of most county FWAGs inevitably devolves upon a smaller group. One FWAG secretary reported "Most FWAG work/decisions undertaken by chairman, secretary, NFU county secretary, county trust conservation officer, ADAS adviser". This group was unusual in not having a formal steering or executive committee. Some 32 out of 38 of the groups had such a committee, which ranged in size between 6 and 18 members with a median of 9. Significantly, the groups without a steering or executive committee were the ones without a full-time adviser. Evidently the appointment and direction of the latter demands more formal and routine decision making within a group, as well as giving it a sharper sense of its separate identity. Typically, whereas the wider committee might convene for business meetings 3 to 4 times a year, the steering committee might meet at least twice as frequently, and in one case claimed to meet every two weeks.

Key Honorary Officers

Ultimately, of course, the character, level of activity and effectiveness of a county FWAG will depend very much on the qualities of its key officers. In interview, Eric Carter, then national FWAG adviser, identified "the energy and personality of the chairman and honorary secretary" as major factors in the success of a county FWAG (personal interview, 8 February 1984). Of the county chairmen of the groups in our survey, 32 (84 per cent) were farmers. The rest included the principal of an agricultural college, an ADAS Divisional Agricultural Officer, two retired ADAS officials, and two land agents. Usually, a chairman is a well-respected and well-connected local farmer or landowner with a long-standing interest in conservation, typically someone who has been or is a senior officer or committee member of the local NFU or CLA and may also have served on the committee of the county trust.

Vice-chairmen are largely of a similar mould. Of the 29 groups with this office, 18 were filled by farmers, and most of the rest included farm managers, land agents and present and former ADAS officials. Significantly in the few cases where a chairman was not a farmer, the vice-chairman was either a farmer or farm manager, reflecting an understandable wish firmly to root the group in the local farming community. However, in as many as half of all the groups, both these offices – the chairmanship and the vice–chairmanship – were filled by farmers, farm managers or land agents. Indeed, only two groups clearly broke with the pattern of having farmers or agriculturalists in these two leadership positions: Northumberland with an NCC regional officer as Vice-Chairman; and Kent with an ecology lecturer as co-Vice-Chairman alongside a farm manager. It might be advisable for other groups to follow this lead and have a prominent farmer/landowner as chairman and a respected conservationist as vice-chairman or co-vice-chairman. Certainly such a combination would be more appropriate to groups that claim to offer a neutral meeting ground between local farming and conservation interests. It would help to improve the credentials of county FWAGs and effectively counter such charges as that made by one county trust in response to the RSNC's survey, that the local FWAG had "a chairman who does not know what conservation is" (quoted in Heaton, 1984). It should also dispel the suspicion aroused by just a few of the county groups that they too closely mimic the local NFU hierarchy.

The smooth running and efficient operation of a group depends very much on its secretary. At the time of our survey one group was without a secretary. Of the remaining 37, 24 (65 per cent) had county secretaries who were ADAS employees – mostly surveyors in the LAWS division but also some field and drainage engineers and a few agricultural advisers. The other 13 county groups had secretaries with a variety of backgrounds: 6 were employed by county councils in such capacities as planning or countryside officers or college lecturers; and the remainder included a farmer's wife, a forester, an accountant and a university lecturer in environmental science. The preponderance of ADAS employees as FWAG secretaries is a development of the late 1970s and subsequently. Prior to 1978, agricultural college lecturers represented the largest group of county secretaries. In our survey, however, only 3 were recorded (including one who had retired from his lecturing post).

The amount of time spent by the county secretaries on FWAG work varied between 1.5 and 15 hours per week, with a median of about 4 hours. Much of this is done in people's spare time. But most FWAG secretaries are also doing part of the work at least as an integral element or extension of their main professional duties. As one FWAG secretary who was an ADAS surveyor commented: "The amount of time spent on FWAG work cannot be accurately assessed as organizing of events is often in conjunction with ADAS anyway". Indeed, only 6 FWAG secretaries were not employed by organizations that are members of FWAG, and of these 6, it is significant that 3 had no professional commitments, being either housewives or retired, and 2 were self-employed. Even so, involvement with FWAG does enjoy the official blessing of ADAS, and it is therefore not surprising that ADAS officers tend to put in more time than do the other county FWAG secretaries. However, one ADAS Rural Estate Surveyor who devoted between 10 and 15 hours per week to the task did complain that this was "too many, both at work and at home".

A secretary's work also tends to be more involved for groups with an adviser than those without. This may seem paradoxical. After all, the appointment of a full-time staff member might be expected to relieve existing work loads, particularly the burden of having to respond to queries and requests from farmers. But additional administration and accounting is needed to maintain and supervise the appointment, and an adviser so boosts a group's activity and profile that work loads can increase all round (see Chapter 7).

Advisers

In recent years, most FWAGs have been able to appoint their first full-time officers, and in a few cases (including two in our survey) two neighbouring counties have combined to support such an appointment. Normally known as farm conservation advisers, the intention in all cases is that they should offer an advisory service for local farmers and landowners. Some 24 full-time advisers were employed by 26 county groups (out of the total of 38 included in our survey), 17 of them having been appointed during the previous 16 months (i.e. since the beginning of 1985). A number of the remaining groups had plans or hopes of appointing an adviser (Figure 4.1).

Figure 5.2 illustrates the geographical coverage of FWAG officers at the beginning of 1986. The Farming and Wildlife Trust hoped to complete the coverage of England and Wales by early 1989 but although this was achieved in England (excepting Staffordshire which actually lost its adviser due to lack of farmer support), in Wales only three counties had advisers by this date. In a few counties there is no intention to appoint a FWAG adviser because conservation advice for farmers is already given by an official of another organization who liaises closely with the FWAG committee – such as the countryside advisers employed by Bedfordshire and Avon County Councils and by the Montgomery Trust for Nature Conservation (all three posts are part funded by the Countryside Commission). The counties where progress has been slow are in the less prosperous farming areas of the west, within the Less Favoured Areas or where the farm structure is dominated by small farmers. Only one Welsh county (Gwynedd) had a FWAG adviser at the time of our survey. In these areas, raising local funds has been the main stumbling block. One county secretary outlined the difficulties as follows: "Finance not yet satisfactorily sorted out – quite a good response from ancillary trades, but response from farmers disappointing. With the drastic reduction in farm profits over the last year or two, conservation is low on the list of priorities".

Being a FWAG adviser is a demanding task, and not just in terms of the work load. Ideally, he or she should have *both* the social skills to win the trust of local farmers and to deal diplomatically with the range of organizations represented on the FWAG committee; and the right technical expertise in order to appreciate both the farming and conservation problems of the area and to give practical advice. These may seem exacting requirements, and in the nature of such a novel activity there is no standard training that would fully equip someone. Indeed, to the extent that the individuals in post are creating the profession of farm conservation adviser *de novo*, they are also gaining much of the relevant experience and skills on the job, although recently there have been efforts more systematically to identify and meet their training needs (Kerby and Hastings, 1987).

The earliest appointments, in the late 1970s, emphasized a working knowledge of farming, and the advisers lacked formal qualifications in conservation or natural history (see Chapter 4). Only two of the advisers in our survey conformed to this model. Both had strong agricultural credentials and farming experience. Although no relevant conservation qualifications were recorded for either of them, one was described as "an amateur naturalist" and the other as "a keen countryman". Even so, this model does seem to place much greater priority on ease of rapport with farmers than on the technical proficiency (from a conservation viewpoint) of any advice given.

All the other advisers in our survey had some conservation experience and/or relevant qualifications in such fields as ecology or environmental sciences. All but two had degrees, and most (15) had post-graduate qualifications. One group of advisers (a quarter of the total) lacked agricultural (or forestry) experience or qualifications. Whereas all the advisers are likely to encounter some difficulties in forging links with the local farming community, it is probably amongst this group that the greatest difficulties could be experienced, particularly in gaining the confidence of farmers not wholeheartedly committed to conservation.

The largest group of advisers (two-thirds of the total) were blessed with *both* a farming background or agricultural (or forestry) qualifications *and* conservation experience or qualifications. For some, the extent of their farming background was having been brought up on a farm, but this was evidently regarded as of sufficient practical or psychological significance for their work as to warrant being reported under the heading "relevant experience and qualifications". Most, though, seemed exceptionally well qualified. One adviser, for example, was the son of a farmer, had a degree in biology and an MSc in Environmental Technology, and had work experience with the NCC, a planning authority, a water authority and on various farms. Another one was a geography graduate with an MSc in forestry and a diploma in landscape design who had worked for seven years as a landscape architect and as a Forest District Manager with the Forestry Commission.

As well as developing their skills on the job and learning from the experience and knowledge of FWAG committee members, the advisers have also benefited from courses mounted under the auspices of national FWAG. The advisers are formally employed by the Farming and Wildlife Trust which reserves ten days of their time each year for national purposes, mainly training but also promotional events. Various sessions and field visits have been mounted often in conjunction with some of the organizations represented on national FWAG, but also with others such as the Game Conservancy, the Institute of Terrestrial Ecology and ICI's Jealotts Hill Research Station. A day organized by the Association of County Archaeological Officers examined the problems of preserving archaeological features on farmland; and a three-day course on woodland management was mounted by the Royal Forestry Society.

These specific events supplemented a flow of information from the national FWAG adviser. A bi-annual newsletter not only keeps the county advisers (and other local FWAG officers) informed of important news and events but also brings to their attention relevant technical reports and provides briefing notes on various agricultural and conservation topics pertinent to their work. Back issues have included descriptions of the role and organization of the British Association for Shooting and Conservation, the Weed Research Organization, ADAS's Wildlife and

Storage Biologists and the British Hedgehog Preservation Society. Advice and explanation have also been given on the use of tree shelters (or Tuley Tubes); on rabbit-proof fencing; on acid rain; on a new attachment to fertilizer distributors to eliminate spreading into hedges; on the benefits and control of ivy; and on the Code of Good Agricultural Practice issued under the *Control of Pollution Act*.

One further attribute of most FWAG advisory officers calls for comment. More than two-thirds of them are women. Evidently, the Farming and Wildlife Trust and the county groups deserve to be congratulated on their progressive hiring. What makes the matter so curious is that farming is such a male-orientated and male-dominated occupation. Indeed, in other respects, FWAG itself conforms to this norm. National FWAG has no women amongst its 29 members and county FWAGs are also largely male bastions. All 64 county chairmen, for example, are men, and only six county secretaries are women. How then are we to understand the preponderance of women in the crucial role of advisory officer? After all, this is now the key executive role, the main point of contact between a FWAG and local farmers, and the one role in the entire FWAG structure that is remunerated (which neatly reverses the traditional relationship in our society of men's work being remunerated and women's work not).

First, let us consider the question of supply. The women advisers in our survey tended to be somewhat better qualified than the men; they were more likely to have *both* a farming background or agricultural qualifications *and* conservation experience or qualifications; and they were more likely to have a post-graduate degree. The graduate courses in conservation, landscape and applied ecology that several of the advisers had taken do attract a lot of female students. There is thus a good supply of well-qualified women for such posts as the FWAG ones. But this does not account for their numerical superiority in these posts as those courses train as many or more men. Another salient factor may be a clear preference for FWAG advisers with a farming background or at least a family background in farming (of the 24 advisers in our survey, at least 11 came from farming families). Farmers' sons who go into higher education intent on a farming or related careers are likely to study agriculture. Farmers' daughters are much more likely to study some other subject and, if strongly interested in nature or the countryside, this might well be geography or biology or environmental studies. There are thus likely to be more suitably qualified farmers' daughters than farmers' sons for the FWAG advisory posts. But this explanation merely begs the wider question of why farmers' sons are so often channelled into an agricultural career and farmers' daughters are not. In addition, half the advisers, including half the female ones, are not from farm families.

This takes us to the demand side, and here it is difficult to escape the conclusion that there may be a prevalent, albeit implicit, assumption that FWAG's message and advice would be better relayed to farmers by a woman than by a man. Just as agricultural production is so closely associated with masculine values, so the non-productive, nurturant and caring activities of the farm household are closely identified with feminine values and the female role. Conservation, it seems, has become associated with the latter. It is now part of the folk wisdom within FWAG circles, for example, that "The farmer's wife can often be a useful ally in getting conservation ideas accepted" (FWAG Newsletter Autumn/Winter 1981/82), and FWAG's very first pamphlet urged farmers to think carefully before undertaking too drastic remodelling of their farms lest they come to "regret the treeless, hedgeless prairie view from the farmhouse windows" ('Farming with Wildlife', paragraph 1). Such remarks would appear to draw upon notions of the feminine as a humanizing presence, constraining and civilizing masculine power and vigour. But they may also carry other connotations as something pleasantly diverting or even fanciful, and certainly removed from the serious (manly) business of wresting a living from the soil.

Advice to Farmers

The main role of the advisory officers is giving conservation advice to local farmers. There are, of course, many other pressures on their time. One officer listed the following: talks to farming and non-farming groups, preparing material for the local press and radio, lecturing to agricultural students, preparing and manning exhibitions, and organizing farm walks. Significantly, a number of FWAG steering committees and officers have had to lay down fairly stringent guidelines to ensure that these other activities do not impinge too much on the advisory work.

The number of farm visits per year by the advisers varied between 66 and 263 (including follow-up visits). The median figure for new farm visits was 100, and a number of county secretaries commented that this was the norm or the target that advisers were expected to reach. Many more requests were dealt with over the phone. One adviser reported 112 telephone consultations with farmers in one year. The publicity given to the appointment of an adviser, the spread of knowledge about her functions and the ready availability of someone to respond stimulate a greatly increased flow of requests for advice. Wiltshire FWAG, for example, received more requests in two months with its adviser in post than in the whole five years of its existence prior to the adviser's appointment. Most groups reported between 20 to 25 requests for advice in the year prior to the appointment of an adviser, and for most this had since risen to at least 100 a year. For those groups in the survey without an adviser, the median figure for requests for advice per year was 20, though several were uncertain or did not know. As one county secretary commented, "it is often difficult to differentiate whether requests are coming to FWAG or to the naturalists trust, ADAS, or the county council". Evidently one consequence of the appointment of an adviser is to crystallize the existence of FWAG as a distinct source of conservation advice.

A key tenet of FWAG's voluntary philosophy is that advice-giving should be responsive. As Jim Hall declared "advice is for those who request it, not to be thrust upon people. The demand, we have proved, develops from awakening interest and setting an example" (FWAG Newsletter, Autumn 1979). At first, inevitably, the spread of information about the availability of FWAG advice depends on the promotional efforts of its member organizations and any publicity the appointment of an adviser attracts. Initially, also, many if not most requests for advice come through the member organizations such as the county trust, the NFU or the County Council. As the adviser settles into the job and knowledge of the scheme spreads, so the pattern of introductions to visits changes. Increasingly, requests originate from farmer-to-farmer contacts, as well as via ADAS referrals, and the direct promotional efforts of the adviser, for example through the farming press, at agricultural shows or at meetings of farmers. As well as giving advice directly, the advisers are also armed with a range of special leaflets prepared by national FWAG (see Appendix).

Table 5.1 indicates the most common requests for advice received by those county FWAGs that employ an adviser. Tree planting was the topic most in demand; perhaps not surprisingly, because, at some scale, this is a feasible option for most farms and there are different grant schemes operated by the Forestry Commission, the Countryside Commission, MAFF and some local authorities. Of the 24 groups that answered the relevant question, 11 ranked tree planting first, and only Hertfordshire, Kent and Northumberland did not include it among their three most popular topics. The second most common request overall was for advice on the management of existing features, with woodland and grassland

Fig. 5.2 Countryside Advisers supported by Countryside Commission, 1986

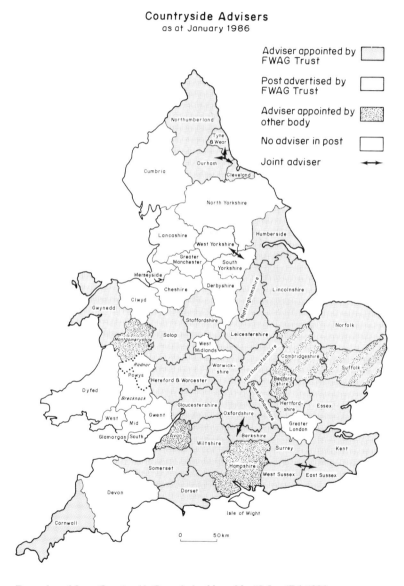

Reproduced from *Countryside Commission News*, No. 19, Jan/Feb 1986.

Table 5.1 Most Common Requests for Advice to County FWAGs with an Advisory Officer

	No. of groups ranking this topic:			Total number of groups (out of 24) giving this topic any ranking (1 to 6)
	1st	2nd	3rd	
Tree planting	11	6	4	24
Management of existing features	7	6	1	22
Pond creation/ restoration	5	9	4	23
General farm survey	2	2	7	20
Hedge Planting	-	2	7	22
Farm trails	-	-	-	6

Other topics mentioned by one or two groups each (but none of them ranked first, second or third) included: creating new wildflower areas; bird or bat boxes; screening; archaeological matters; caravan sites; reed beds and marginal water plants; establishing low maintenance grassland; and sources of labour.

Note: the columns do not add up to 24 because a few responses gave some topics equal rankings or did not use all the rankings.

management being the most prominent followed (in declining order) by ditches and dykes, hedges, ponds, and heathland and moorland. Close after this topic, in third place, came pond creation. A number of groups ranked it first, including Cornwall, Dorset, Leicestershire; but for a few, including Kent and Oxfordshire, it had a very low profile. A somewhat lower priority was general farm survey work, often as part of, or the prelude to the preparation of a farm conservation plan. This came fourth overall, though Herefordshire and North Yorkshire placed it second, but for a number of groups, including Cornwall and Leicestershire, it had little if any significance. Finally, several groups identified the establishment of farm trails, besides a number of other topics (see Table 5.1) as significant subjects for advice, though not the most common ones.

The results of the CRS survey of 113 farmers carried out in six counties (Lincolnshire, North York Moors, Bedfordshire, Somerset, Gwynedd and Tyne Tees) in 1988-89 are consistent with these findings. Forty-seven per cent of farmers had sought FWAG or, in the case of Bedfordshire and North York Moors, local authority farm conservation advisers' advice for the creation of new features, 37 per cent for new features and management of existing features and only 16 per cent solely for the management of existing features. Advice concerning established features was slightly more likely to be implemented (in 72 per cent of cases) than advice on new features (68 per cent) (CRS, 1990).

It is difficult to discern an obvious pattern or logic in many of these variations. Thus, while it is not surprising that hedge planting is a minor issue in a county such as Cornwall, which has not witnessed a greal deal of hedgerow loss and where hedges are not a major landscape component, it is more difficult to understand why it should be of such significance in Herefordshire but of so little importance in Essex. One can only conclude that much depends upon the priorities expressed or identified by the local FWAG committee and its adviser. One pattern is discernible, though. The longer a FWAG adviser has been in post the greater the priority that tends to be given to management and/or survey work. Conversely, the more recent the adviser's appointment the more likely is higher priority to be given to hedge planting, tree planting and pond creation.

There is considerable pressure on new FWAG advisers to demonstrate immediate and tangible results. Not only may the creation of new features satisfy the individual farmer or landowner but in gross terms as, say, the number of new trees planted, it also yields statistical evidence of achievement, to be reported to sponsors and overseeing committees. From within national FWAG itself, however, concern has been expressed over the possibility that "many farmers believe that the planting of trees in field corners is a substitute for the disappearance of pieces of old woodland, ancient grassland, or wetland sites on farms" (FWAG Newsletter, Spring 1980; see also Information Leaflet No 2 – 'Wildlife on the Farm: FWAG Guide to Priorities'). A number of sources, including the farming media and ADAS conservation training courses, have tended to give disproportionate emphasis to the creation of new features (Corrie, 1984, pp. 41, 86 and 140). With time, the advisory officers, with their conservation training, may be able to impress on their farmer-clients and their local committees, if necessary, the much greater value of retaining and sympathetically managing areas of natural and semi-natural habitat than creating habitats *de novo*. Often only through an invitation on to a farm to discuss a proposed planting or drainage scheme will existing habitats and the threats they face come to light. Of course, on a farm that has lost its original habitat the making of new plantations and ponds may be worthwhile to restore at least a few of the species that have been lost and improve the landscape. But such efforts should not be allowed to eclipse the vital work of bringing to the attention of farmers what is already of value on their farms from a conservation viewpoint and what steps should be taken to safeguard and enhance its value as an integral part of their farm management.

The evidence we have concerning requests for advice received by those county FWAGs without an adviser (see Table 5.2) seems to confirm the tendency for greater priority for management of existing features and farm-survey work to follow from the appointment of an adviser and the ecological perspective she can bring to bear on the outlook of a county FWAG. Thus, whereas tree planting also predominates for these adviser-less groups, in contrast pond creation and hedge planting come second and third as common requests for advice; and farm survey work and the management of existing features trail way behind in fourth and fifth place.

Table 5.2 Most Common Requests for Advice to County FWAGs without an Advisory Officer

	No. of groups ranking this topic:			Total number of groups (out of 11) giving this topic any ranking (1 to 6)
	1st	2nd	3rd	
Tree planting	9	2	-	11
Pond creation/ restoration	4	5	2	11
Hedge planting	1	3	4	10
General farm survey	-	1	1	8
Management of existing features	-	-	4	11

Note: the first column does not add up to 11 because a few responses gave two topics equal rankings.

Those advisers interviewed in the CRS study claimed they gave greatest priority to the management of existing features, but they continued to experience considerable resistance to this emphasis from the farmers they visited, who often wished to undertake a new project. One adviser estimated that he advised the management of existing features, as against creating new ones, at a ratio of 3 : 1 but that clients actually implemented elements of his advice in inverse proportion.

These findings concerning the impact of the appointment of advisers on the priorities of county FWAGs go some way to dispel the concerns of

The Workings of County FWAGs 103

conservationists, such as those expressed by some county trusts in the survey conducted in 1984. A sizeable minority (10 out of the 39 trusts surveyed) stated that it was important that FWAGs should place more emphasis on habitat conservation and less on what was seen as the simply cosmetic work of tree planting and the like, described variously as "the field corner syndrome" and "tarting-up farms". Indeed one trust stated that there was a great danger in making FWAGs more effective if they merely reinforced this approach (Heaton, 1984).

Though advice given to farmers on an individual basis is becoming the primary activity of county FWAGs, they continue to mount a variety of other activities mainly aimed at stimulating the interest and involvement of the local farming community as well as maintaining an interchange of ideas and information between agriculturalists, conservationists and landowning interests. Table 5.3 indicates the range of meetings and events organized by county FWAGs during 1985. Farm walks, visits and open days were the most common events, followed by talks and discussions – the former being the favoured event of the spring and summer; the latter, of the autumn and winter. Appearance at an agricultural show was the other prevalent event, reported by at least a third of the groups. Other activities mounted by some of the groups included conferences and field demonstrations, social events, farming and wildlife competitions, and fund-raising events. A minority of groups (10) arrange some of their events in conjunction with another organization, most usually ADAS.

While Kent and Sussex reported 32 and 25 events respectively for the year, Ceredigion reported none and Montgomery just one. Not surprisingly, the more active groups tended to be those with advisers. The median figure for the number of events they held was 7, compared with 4 for those groups without an adviser in post by 1985. The most energetic groups tended to be those with an associate or open membership scheme: amongst them the median figure for the number of events was 10.

Of the 38 groups in our survey, 15 had some sort of membership scheme, all but two of these being groups with an adviser. By and large, these schemes are intended as a means of associating sympathetic farmers and landowners with the work of the county FWAG. Membership numbers varied from 5 for a group that had just recently launched its scheme to the 400 supporters of Sussex FWAG, with most groups in the range between 70 and 170. Typical subscription rates were £5 to 10 per annum and for these supporters usually received an occasional newsletter. Some FWAGs also look upon their memberships as potential sources of donations to support their work: Worcestershire FWAG, for example, reserves the title

Table 5.3   Non-Business Meetings and Events Organized by County FWAGs during 1985

Type of event	No. of groups that specified the event	No. of events reported to have been held
Farm walks, visits and open days	26	73
Talks and discussions	14	68
Appearances at shows	13	43
Conferences/field demonstrations	10	13
Social events	8	11
Competitions	8	11
Fund raising/PR events	5	5
Total no. of groups that provided usable answers	36	
Total no. of events reported by all these groups		281

Note: the figures do not add up as some groups did not provide a breakdown of the types of events they had held, simply the overall number.

of patron for those members who covenant at least £50. Of the 23 groups without a membership, three were contemplating launching their own schemes in the near future. The other groups, though, had either no need or had rejected the idea on various grounds: such as, that a membership scheme would be cumbersome to operate and maintain; that it would make the group too closed; that it might deter large subscribers or donors; or that members might expect to have some control over the group.

Effectiveness of Advice

The study conducted by the Centre for Rural Studies sought to establish the effectiveness of advice on the 113 farms visited through questions to the farmers and through field-survey work on each holding. In each case a judgement was made on whether the advice had been broadly implemented, partially implemented or not implemented at all. Table 5.4 shows the results of the survey. The investigation was carried out in the

Table 5.4 Number and Percentage of Farmers Implementing Advice by Region

	Broadly Implemented	Partially Implemented	Not Implemented
Lincolnshire	8 (44%)	3 (16%)	7 (39%)
North York Moors	4 (24%)	9 (53%)	4 (24%)
Bedfordshire	15 (75%)	1 (5%)	4 (20%)
Somerset	12 (60%)	1 (5%)	7 (35%)
Gwynedd	7 (35%)	4 (20%)	9 (45%)
Tyne Tees	8 (44%)	4 (22%)	6 (33%)
Total	54 (47%)	22 (19%)	37 (34%)

autumn and winter of 1988-89 and referred to advice offered between April 1986 and March 1987.

Just under one half of the advice is broadly implemented and a third not implemented at all. What is perhaps noteworthy are the regional disparities, with the lowest levels of failure in Bedfordshire and North York Moors, where the advisers were employed not by FWAG but by the respective county council. Despite a certain antipathy amongst some of those active in the development of FWAG towards county councils assuming a leading role in the provision of farm conservation advice, advisers employed by local authorities are able to draw on an extensive range of support services to back up their advice. FWAG advisers do not usually have direct access to such support and therefore their advice tends to be somewhat less effective.

The CRS team also assessed the quality of the advice given through judgements of both the farmers' perceptions of the advice and field assessments. In the estimation of both farmers and researchers the advice given was of a high quality, over 80 per cent scoring 4 to 5 on a scale from 1 to 5. However the team concluded that too little attention was given to the possibilities for whole farm plans. Issues of landscape and public access, as opposed to wildlife, were widely neglected. Nonetheless, the overall view was that FWAG advice was of a generally high quality and achieved moderately high levels of effectiveness, but with considerable regional and inter-farm variations. The failures of FWAG advice largely reflected wider failures in agricultural and conservation policy, and also the weak institutional support for the advisers' work.

Conclusions

The survey results present a picture of county FWAGs in transition: indeed their organization and functions continue to evolve. It is also evident that the groups vary to a considerable degree, not only in their composition (reflecting, no doubt, differences in the structure of farming and rural politics between counties), but also in the scale and nature of their activities. Nevertheless, it is possible to draw a few broad conclusions particularly regarding some of the ingredients essential to their effectiveness.

A first requirement for a county FWAG is the commitment of a core group of individuals with the necessary leadership and organizational qualities, good connections within the county and a genuine interest in conserving the farmed and wooded countryside. Second, particularly through its membership but more generally, a county FWAG needs to be well integrated into the local network of organizations dealing with agriculture, forestry and conservation, and to have good working relationships with them. Third, it is evident that the appointment of full-time advisers has dramatically changed the scope of county FWAGs. Many have been transformed from small cosy coteries into energetic, outward-looking and professional groups. Our survey results also suggest a shift in outlook and the infusion of a more ecological perspective, following the appointment of an adviser, leading to a greater emphasis on the identification, rehabilitation and management of established habitats rather than the creation of new features which can too easily be stigmatized as cosmetic.

Of course, the effect of these efforts will depend on the responsiveness of the local farming community as well as the receptiveness of individual farmers and landowners to conservation advice. It is this which, in the final analysis, will be the crucial factor determining the potential effectiveness of county FWAGs for, without broad local support, their impact will be circumscribed as will their ability to sustain a professional advisory service. In the next two chapters, we explore in detail how these and other factors have interacted in two contrasting areas.

CHAPTER 6

Stony Ground: FWAG in Wales

Introduction

There are compelling reasons for giving special attention to developments in Wales. For a start all the Welsh county groups are known as Farming, Forestry and Wildlife Advisory Groups (and are hence known as FFWAGs). The pace and emphasis of these initiatives have been in contrast to those in England, and national FWAG at Sandy has played a less direct role. This chapter presents both a general account of the development of FFWAG in Wales and a specific account of the Montgomery group (the first to be formed in the principality) including a survey of its members' views and attitudes.

The distinctive history of the Welsh FFWAGs has to be understood in the context of other differences between countryside politics in England and Wales. Many of these stem from the formation in 1964 of the Welsh Office, which expanded its remit somewhat slowly in the 1960s and more rapidly in the 1970s. It took some responsibilities for agriculture in 1969, but MAFF relinquished control reluctantly (Rowlands, 1972) and the process was not completed (if indeed it can be said to be completed yet) until 1978, when the Secretary of State for Wales assumed responsibility for ADAS. Even so MAFF continues to supply the ADAS staff to the Welsh Office Agriculture Department (WOAD). So too the Forestry Commission's activities in Wales are nominally overseen by the Welsh Office; whereas the Countryside Commission and the NCC had separate Welsh Committees before the establishment of a united conservation agency for Wales.

The NFU, while operating with its usual county and branch structure, also has a Welsh Council formed (as the Welsh Committee) in 1963, largely as a response to the growth of the Farmers' Union of Wales (FUW). The Council is responsible for co-ordinating the Welsh viewpoint but remains under the executive authority of the Union's national Council. It has its own secretariat responsible for implementing Union policy in Wales, but is heavily dependent on the professional services of London headquarters. In conservation issues the Land Use Division of the Parliamentary Committee, based in London, has taken the lead.

Prior to the formation of the FUW in 1955, the NFU had a representational monopoly of farmers in England and Wales. The formation of the FUW stemmed from the dissatisfaction of hill sheep farmers and marginal dairy farmers with an NFU perceived to be dominated by big lowland arable farmers. The competition between the FUW and NFU has been intense and remains apparent in attempts to outdo each other in the defence of farmers' interests, not least in response to conservation pressures. However, Welsh political opinion has been less exercised on conservation issues than metropolitan opinion. Arguably also, the NFU has been more alert to the rising popular interest in conservation and anxious to present itself nationally as sympathetic to 'sensible' conservation. On the other hand, the FUW as a more marginal and radical group without the corporatist history of the NFU, is more single-minded, some would say parochial, in the pursuit of narrowly defined farming interests.

The FUW and Conservation

It has to be said, of course, that unlike large tracts of the English lowlands much of rural Wales is of national conservation importance. There are five Areas of Outstanding Natural Beauty (Lleyn, Anglesey, Gower, Wye Valley and Clwydian Range), over 700 Sites of Special Scientific Interest, and three National Parks: the Brecon Beacons, Snowdonia and the Pembrokeshire Coast. In a number of localities the FUW has been actively involved in opposing designations. In the early 1970s it successfully resisted the designation of another National Park in the Cambrian Mountains, and it has consistently opposed the formation of an AONB on the Berwyns. It failed, however, in a long campaign against the Clwydian Range AONB and has had limited successes in campaigns against a number of SSSI designations.

The Union has seen such designations, in nationalistic terms, as English impositions designed to cater for English tourists, and has contrasted the treatment of Wales with Scotland where, it argues, a greater degree of national autonomy has ensured that the Highlands have not become similarly encumbered. Thus the FUW has been engaged in conservation politics, broadly defined, almost since its inception in 1955, but its stance has tended always to be reactive. Long-term issues such as habitat protection and the integration of farm management and conservation of wildlife have only recently commanded any attention. Indeed, the rhetoric of the Union's public utterances on a number of these matters has done little to nurture the spirit of compromise and goodwill so assiduously cultivated by the NFU. In fact, the FUW's attacks on the conservation organizations, particularly the Countryside Commission, have at times been little short of vituperative.

More damaging to the Union's reputation within policy circles was its rejection, in 1977, of an invitation from the Welsh Committee of the Countryside Commission to join the NFU and CLA in drafting a code of practice for farmers aimed at improving the integration of farming and conservation (see pages 39-40). The Union roundly declared that no such code was needed and that good husbandry should be followed not preservation. In similar vein the Union opposed the Strutt Report's recommendation that ADAS should be empowered to give conservation advice, arguing instead that ADAS should only be involved in improving the efficiency of farm businesses and raising levels of farm income.

The Union has plainly felt perplexed by the mounting challenges to farmers' freedom of action, particularly during the protracted battle over the designation of the Berwyn SSSI (1978-1982) and the passage and implementation of the 1981 *Wildlife and Countryside Act*. The plea again and again has been that farming is a business and should be allowed to develop as it sees fit: the farmer's role is to produce the nation's food not to act as a countryside warden. Not surprisingly the Union has been unreceptive to attempts, such as those provided by Countryside Commission demonstration farms and FFWAG, to suggest that compromise solutions are technically and economically feasible. Thus the FUW's involvement in both FFWAG and the Wales Countryside Forum (see below) was premised at least initially on the need to counter undesirable conservation initiatives. A few members did recognize a wider rationale. As one senior FUW member commented in an interview:

> I spoke very strongly in favour of involvement at the beginning. I knew that the basic principle was right: small-scale, non-contentious, education – we all need to learn — low-key, local, simple, straightforward, not revolutionary. Also to be cynical it is a way of turning round to the conservationists and saying 'you can't accuse us of not being interested in conservation'. Some of the generally more politically minded farmers have an eye to that. They think it's worth getting involved in something like this.

As individual FUW members have thus become involved some have become progressively less critical of conservation and more conciliatory. The extent to which they are able to carry the Union's full authority may, of course, be limited.

These observations suggest strongly some of the reasons for the distinctive development of FFWAG in Wales. In particular two competitive farming lobbies have had to work together and their rivalry has hardly assisted the development of FFWAGs. Locally, neither organization could contemplate finding itself in a position where the other could portray it as being 'soft' on conservation. It would, however, be a mistake to see this as the sole reason for FFWAG's slow growth in Wales. The particular farming problems of the uplands and alternative initiatives to reconcile farming and conservation have both also had an impact.

The Dinas Conference: A False Start

As early as 1970 NAAS and the Montgomeryshire Countryside in 1970 Committee organized a conservation farm walk at Castle Caerinion, on a farm which subsequently became a Countryside Commission demonstration farm. It belonged to R.O. Hughes, a former Deputy President of the FUW and Chairman of the Union's Policy Group. Hughes developed his interests in conservation largely as a result of representing the Union on the Prince of Wales Committee, a loosely structured environmental forum set up in the 1960s under the auspices of the Countryside in 1970 Conferences.

Subsequently Hughes represented the Union in discussions concerning the formation of FFWAGs in Wales. His role in organizing the farm conservation walk in 1970 prompted full coverage of the event in the FUW's journal, *Y Tir*. Two years later, however, *Y Tir* made only passing reference to a far more significant event which took place in Carmarthenshire.

The Dinas Conference, held in August 1972, was a clear attempt to carry into Wales the impetus created in England by the Silsoe Exercise. It was organized by FWAG in conjunction with the Forestry Commission and ADAS. David Lea, Eric Carter and Derek Barber enlisted Ronald Soden, the Regional Officer for ADAS in Wales, to chair the organizing committee. During the early 1970s forestry was seen as the major force for change in the uplands, and hill-farming practices were not seen to present such a challenge to conservation. Hence the potential for forestry development was a major focus of the conference, and conflicts between agriculture and forestry were given far more weight than farming and wildlife matters.

The Conference was opened by the Director-General of ADAS, Sir Emrys Jones. In his introductory talk he made no mention of 'wildlife' or 'conservation', but called for an 'ecological' survey whose main purpose would be to assess the suitability of land for forestry (MAFF, 1975, p.9). Sir Emrys catalogued the increasing agricultural productivity of the hills and ways in which further improvements, for example to rough grazings, could be made and were likely with Britain's advent to the Common

Agricultural Policy. His concluding remarks displayed little awareness of the likely reaction of conservationists to plans for wholesale improvements to upland agriculture:

> The Community (EEC) policy towards hills and uplands is to inject capital into them and to try and preserve reasonable social rural communities that are viable by various means. What does this mean for the Towy Valley? This is going to mean certain physical changes: more fences, better access, less *Molinia*, more *Agrostis*. It is going to mean a change in colour. It is not going to mean a lot of wholesale ploughing. It is not going to mean the cutting down of trees, hedges and that sort of thing, but it is going to mean changes to the landscape. The question is – how do we do it? I believe you really go for a viable agriculture as the basis, then work out the ecological aspect, or land environments you want to see 20 years from now. I am certain that there is little conflict between the two if you are quite clear what you are aiming for. I am quite convinced that the most important aim is a viable rural community with its own social activities and its own aspirations. (MAFF, 1975, p.9)

Following the pattern of the Silsoe Exercise, Dinas took the form of syndicate groups making farm plans with different sets of objectives. The *farming syndicate* was asked to produce a plan for the maximization of income on the 1200-ha hill farm. The *farming and forestry syndicate* was instructed to release 240 ha for commercial forestry and to maximize farm profits on the remaining land. A *forestry syndicate* was to produce an investment syndicate plan for total afforestation. In the first three syndicates the participants were asked to take into account payment of Estate Duty, whereas in the fourth, the *wildlife syndicate*, no such condition was made.

For the *wildlife syndicate* the strict business objectives were replaced by the following instructions:

> Your whole existence centres round your love for the Welsh hills and their wildlife in which you have a broad interest and know well. The farm has to provide you with sufficient income for you and your wife to live including the provision of a television and a small car, but you have no further ambitions . . . . Prepare a plan indicating the management of the land to allow for as much diversity of natural interest as possible within the above limitations and to include some

afforestation so that you can study and experiment with the resulting changes in wildlife. (MAFF, 1975, pp.28-29)

It is surely revealing that the wildlife case is not only limited to some kind of special, non-farming interest, but is even specifically associated with low standards of living! As one of the conference participants put it, in dicussion:

> From the briefs, the farming syndicate and the forestry syndicate have been asked to make money and the wildlife syndicate has been given money to spend, although I think that wildlife might in some circumstances be a money spinner through the tourism aspect. If approached in this way its remit could have been made more directly comparable with the forestry and farming syndicates. (MAFF, 1975, p.45)

The effects of the syndicate plans on wildlife were discussed by Tom Pritchard of the NCC, and predictably he highlighted the detrimental implications for wildlife of the farming and forestry solutions. In response, Meredydd Roberts, Director of Pwllpeiran, the Ministry's experimental farm near Aberystwyth, put forward a 'compromise plan', which involved releasing 200 ha for afforestation and improving more land than was actually proposed by the farming syndicate, although the acreage to be ploughed was reduced from 85 to 60 ha. Shelter-belt planting, amounting to 30 ha, was proposed with the claim that this would provide a positive benefit for wildlife. However, the only proposal of immediate direct benefit to conservation was that of fencing 80 ha of natural oak woodland to allow regeneration.

It is interesting to note that Roberts, an agriculturalist, while accepting the farming and forestry case disputed some of the points offered by the wildlife syndicate. Without offering supporting evidence, for example, he claimed that pasture improvement without ploughing would trigger off 'biological activity which would support more wildlife rather than less'. Indeed the real implications for wildlife were the subject of pious hopes rather than systematic analysis:

> The compromise plan is obviously not the ideal answer to each individual interest; the farmer would prefer not to sell any appreciable acreage for afforestation, the forester would prefer to plant a much larger acreage and wildlife interests would prefer to

have a less intensive system of farming and a smaller acreage of conifers. It must also be admitted that the plan could well result in some changes in the number, distribution and diversity of wildlife in the area, although one would hope they would not be too serious. It is, however, the contention that an efficient and progressive hill farmer is in a much better financial position than anyone else to contribute to the legitimate rights of other interests. (MAFF, 1975, p.38)

In summing up the conference Meuric Rees, a former Chairman of the NFU's Welsh Committee and a future Chairman of the Countryside Commission's Welsh Committee, reiterated the principle of compromise based on priority for agriculture. He hoped that Dinas had provided a blueprint for the whole of Wales. The aims of the Dinas Exercise were clearly wider than those of Silsoe and had policy implications beyond those for wildlife. The question of rural depopulation, for example, was a theme to emerge. But though a holistic approach to the problems is in many ways commendable it is evident that the sense of direction after Dinas was not as clear as after Silsoe, and there was certainly not the same immediacy of purpose.

From Dinas to FFWAG

Nevertheless, the initiative was maintained in a series of informal discussions chiefly prompted by ADAS and the RSPB. In June 1973, a meeting was held in Shrewsbury to consider the establishment of a FWAG in Wales. Another meeting took place in Aberystwyth the following September. As a result a paper was circulated to other conservation and farming organizations proposing the establishment of a Welsh group with objectives and terms of reference parallel to those of English FWAG, including the appointment of a full-time adviser for Wales. Hence a major concern was the raising of money to provide a salary, minimal clerical support and accommodation for the officer.

Wales, however, lacked the financially strong voluntary conservation bodies based in England and, though the RSPB was prepared to back the project, other promises of significant financial support were not forthcoming. So it had to be postponed indefinitely. Efforts over the next few years to revive the initiative came to nothing until September 1978 when WOAD and the NCC collaborated in drafting a constitution for a Wales FFWAG. Both bodies agreed that a central group was the first priority, and the NCC agreed to provide £8000 over three years. However, finance was not forthcoming from the Countryside Commission or the Economic Forestry Group, the two other agencies which might have been expected to provide funds.

In 1980 Soden reluctantly changed strategy and looked towards the formation of county groups instead. After his retirement this strategy was carried forward by his successor as the Divisional Officer for ADAS in Wales, Bill Ratcliffe. Arguably the preoccupation with forming a national body, with a full-time officer, effectively postponed for eight years the eventual attempt at a solution to the Welsh problem which was the piecemeal formation of county groups from 1980 onwards. The first group to be formed was Montgomery, chosen largely at the instigation of ADAS in April 1980, followed closely by Pembrokeshire and Carmarthenshire. The rest of the Welsh county groups were formed over the following 2 to 3 years, and only Radnorshire remains without one.

The exact status of the Welsh FFWAG is not entirely clear. They are affiliated to English FWAG and most have close links with the FWAG headquarters at Sandy and more recently Stoneleigh (as indeed do the Scottish counties). The three advisers appointed to date in Wales, for Gwent, Gwynedd and Ceredigion, are partly financed by the Farming and Wildlife Trust. On the other hand Welsh representation on national FWAG has been confined to Bill Ratcliffe, now in retirement, who until recently (1988) held an unofficial watching brief for FFWAG in Wales. A serious consequence of this, in view of the variety of independent or semi-independent institutions in Wales, is that a number of Welsh bodies have no direct representation on national FWAG. The FUW is the most striking example.

Ratcliffe's work for FWAG was voluntary and unpaid. He had no remit or desire to act as either a troubleshooter or in a promotional capacity, but to help and encourage where possible and when invited to do so. He saw himself as Eric Carter's representative in Wales. The other corporate embodiment of Welsh FFWAGs is the annual meeting of chairmen and secretaries with the national FWAG Chairman and Adviser. As Ratcliffe explained in an interview:

> This is the Welsh FWAG. It is regular. It is a forum and a means of communication from Wales to the national FWAG. I wouldn't subscribe to having a secretariat. The new full-time advisers will come to the meeting, give a snow-balling effect I hope. Representation will widen: we're playing it by ear.

But clearly this forum is very different from the English one which brings together representatives of conservation and agricultural agencies, and it cannot hope to fulfil the latter's mediating role. Another forum exists in Wales, however, and it does fulfil some of the purposes which might have been expected of a national FWAG. The Wales Countryside Forum is a significant attempt to bring together the various farming and conservation agencies in Wales; moreover, its formation was arguably rooted in the failure to establish a fully representative Wales FFWAG.

The Wales Countryside Forum

The Wales Countryside Forum (WCF) was the brainchild of the Countryside Commission. Its genesis was in 1980-81 when the newly appointed director of the Countryside Commission in Wales, Martin Fitton, sought ways of reconciling farming and conservation agencies following the ill feeling generated by the Berwyn controversy. In discussing its programme for the 1980s, the Commission's Welsh Committee was convinced that improving relations was a priority. Informal meetings were held between the Commission and leaders of the NFU, CLA and FUW, which prepared the ground for a 'Conservation Co-ordination' meeting, called by the Countryside Commission, in December 1981, at which twenty-five conservation and agricultural bodies were represented. In summing up the conference, the Chairman of the Commission's Welsh Committee, James Kegie, highlighted three areas of agreement which clearly have close parallels with the objectives of FWAG:

(a) the need for more consultation and co-operation;
(b) the need for farming and conservation bodies to be better informed of each other's needs and problems;
(c) the importance of providing more specialized conservation advice especially through ADAS and the statutory bodies.

The conference agreed to pursue the first two of these items through the establishment of the Wales Countryside Forum with two representatives from each of the main organizations, making around forty in all. The manner in which it has pursued its aims has been quite different from English FWAG in that it has chosen to concentrate at different meetings on a number of major themes. For example, the first meeting in March 1982 focused on AONBs and SSSIs with important statements from the Countryside Commission and the NCC.

Subsequent meetings have been held on agriculture and conservation, access to the countryside, common land and environmental education. More recently the Forum has moved beyond set-piece discussions to deal with issues of the day and to act as an initiatory body. The most notable example to date is the Forum's involvement in promoting woodland conservation. The formation in 1984 of a Welsh Woodlands Group with a secretariat provided by the Countryside Commission, the NCC and the Forestry Commission was a direct, practical off-shoot of the Forum's work. The group's impressive review of the state of Welsh woodlands culminated in 1985 in a highly publicized campaign, 'Save Welsh Woodlands', urging farmers and landowners to seek appropriate grant aid and advice on improving woodland management. The campaign was based on the active co-operation of most of the member bodies of the Wales Countryside Forum, and out of it has emerged 'Coed Cymru' an organization which offers advice to landowners and occupiers on managing and planting woodland. With the financial support of the Countryside Commission, Coed Cymru has now appointed eleven project officers, attached to local authorities, to take the message to farmers in each county in the principality. Its success only highlights FWAG's continuing weakness.

Montgomery FFWAG

When the decision was taken to abandon attempts to form a Welsh FFWAG in favour of the gradual formation of county FFWAGs it was suggested to Soden by the Deputy Regional Surveyor, William Machin, that the old county of Montgomery would be a suitable candidate. He highlighted four reasons for the choice. First, there was the wide variety of land type and farming systems, and second, the existence of a Countryside Commission Demonstration Farm in the county. Third, the ADAS Divisional Surveyor for Montgomery, Ronald Webb, represented Wales on the national Wildlife and Conservation Group established by ADAS's Land and Water Service (LAWS) and was therefore likely to have close contacts with English FWAGs. Fourth, an Area Agricultural Officer stationed at Newtown was particularly enthusiastic. These reasons clearly show the importance of ADAS in making the necessary moves to form county groups in Wales and in lending enthusiastic staff and secretarial resources.

Soden put this suggestion to Tom Pritchard, Director of the NCC in Wales, in December 1979. In April 1980 an organizing meeting was convened by Webb with invitations sent to ADAS, the CPRW, Countryside Commission, CLA, NFU, FUW, Forestry Commission, NCC, RSPB, Timber Growers Organisation, Powys County Council Planning Department, and the North Wales Naturalists' Trust. The meeting resolved that a Montgomery FFWAG should be formed, and Jim Hall was invited to address the inaugural meeting in June 1980.

The successful development of any county FWAG depends upon the degree of internal strength and unity it can establish. In the case of Montgomery there can be little doubt that the concept of FFWAG has been difficult to promote, and its progress has relied too heavily on the commitment of one or two individuals. Most of its early energy was devoted to developing its farmer membership: an obvious priority given that only one farmer attended the inaugural meeting. The group's activities during the first year were relatively encouraging though. Some useful discussions were held at meetings, publicity was sought, farm walks were held and a number of farmers were invited to join the Group. It was even proposed that a Friends of FFWAG be formed specifically for farmers.

These and other initiatives, however, elicited a poor response. One problem was a lack of commitment by the supporting organizations. The FUW, for example, gave little publicity to the group and the CLA refused to contribute to its funds until it had "proved its worth". The group's funding was perilously fragile, dependent in the first year on small grants (£20 to 25) from the RSPB, Countryside Commission, Powys County Council and the NFU. Montgomery FFWAG's Secretary, David Howatson, a local ADAS officer, replied to a MAFF enquiry into the work of County FWAGs carried out in February 1981 that the group "would be disbanded tomorrow if it was not for the encouragement and support given by ADAS". Of all the other member organizations only the RSBP was mentioned as being enthusiastic. Even ADAS could not be relied on to support FFWAG in every way. For example, in 1981, during Montgomery FFWAG's crucial first year, no room was found in the ADAS tent at the Royal Welsh Show for a FFWAG stand. The following year, amends were made when conservation was the theme, but subsequently FFWAG was once more relegated to the forestry tent where it was far less likely to attract farmers' attention.

Undoubtedly, the climate of opinion in the county was not conducive to a co-operative spirit between farmers and conservationists. In particular, the Berwyn controversy cast a pall over Montgomery FFWAG's early years. The NCC, pressed by the RSPB, was engaged in a protracted struggle there over the designation of an extensive upland area as an SSSI which local farmers feared would restrict their opportunities to benefit from afforestation and grassland improvements (see Chapter 9 in Lowe, *et al.*, 1986). Not long after its birth Jim Hall suggested that Montgomery FFWAG should take the lead in bringing the parties in the controversy together. In a letter to Webb, a committee member of the new FFWAG, he described the incident as a "heaven-sent opportunity to show the value of getting people together around a table whose interests seem to be against each other". However, Webb realized that any FFWAG initiative would be too closely identified with ADAS and it could compromise ADAS's position. To invite the NCC's Tom Pritchard into the county to talk about the Berwyns might prejudice his own regular biannual meetings with representatives from the NFU, CLA and FUW (it should be remembered that this incident pre-dated the formation of the Wales Countryside Forum). The issue was subsequently raised at a FFWAG meeting and Webb's line of non-involvement in the Berwyn controversy was endorsed.

Against this unpromising background Montgomery FFWAG suffered, in its second year of operation, the loss of its two key activists, with the departure of its secretary, David Howatson, to another area ADAS office, and the death of its dynamic farmer Chairman, Stephen Williams. The group immediately lost momentum. An executive committee of six members appointed in February 1981 met only once, two months later. No one, it seemed, was prepared to assume Williams's mantle with the same commitment and enthusiasm. The Vice-Chairman, Tom Watkins, reluctantly took over and, by his own admission, did little more than keep the organization ticking over during the next two years. Montgomery's story is by no means unique in revealing the vital role of key individuals in the successful development of county FWAGs. For example, the Sussex group, originally formed in December 1975, lapsed for over a year after the death of its first chairman.

Montgomery FFWAG continues to struggle on. It has had a number of useful farm walks which have provided information for members, but its activities *vis-à-vis* the wider farming community have been minimal. Undoubtedly the development of an active advisory service, prior to the appointment of an advisory officer, presents great difficulty for any county FWAG and Montgomery has been no exception. Few members possessed the time, or probably the skills, to devote to advisory work. In most cases the Secretary (first David Howatson and then Colin Small) has provided what advice has been sought. Howatson reported to the 1981 MAFF enquiry that since the Montgomery group's instigation in June 1980 there had been one formal request for advice, dealt with by the NCC, two informal requests, and a request for a speaker for a local WI meeting. A Countryside Commission survey of all the Welsh county groups' activities in 1984 recorded just 2 or 3 advice cases in the county for that year, compared with 30 cases in Pembrokeshire, 15 in Gwynedd, 12 to 15 in Glamorgan, and 6 in Carmarthenshire.

The severity of the Group's problems was exacerbated early in 1986 when the ADAS Secretary, a keen and knowledgeable naturalist, Colin Small felt compelled to resign. Small was one of the few FWAG secretaries to have belonged to the Agriculture Service of ADAS as opposed to the Land and Water Service (LAWS). While more examples of this might have been applauded by conservationists it raised problems with ADAS where LAWS was supposed to be the lead service for conservation. Hopefully, the amalgamation of LAWS and the Agriculture Service, effected in November 1986, will have overcome such difficulties. In Small's case, however, all work devoted to conservation had been very much by the 'goodwill' of the Agriculture Service which had come under strain with administrative cuts. In addition to his work for FFWAG he was (and continued to be) Secretary of the Montgomery Trust for Nature Conservation. This is a spare time activity to which Small devotes considerable energy. In view of the fact that conservation was restricted to 5 per cent of his work time at ADAS – he wistfully compared this to the 25 per cent allowed in some counties – it is scarcely surprising that he decided to concentrate his energy on developing the work of the Trust.

In consequence LAWS was then called upon again to provide the secretariat for Montgomery FFWAG. But this was not an easy matter, for LAWS had other substantial commitments besides conservation, including drainage, farm buildings and so forth, and had also suffered cuts in its establishment. The LAWS division of the ADAS region covering Montgomery already provided the secretariat to Brecknock FFWAG, although this was hardly an arduous task as the group was inactive. Moreover efforts to form a Radnorshire FFWAG hitherto, had failed: not least because LAWS was unable to sustain a third county group within the region.

The reasons for Montgomery's difficulties are complex. Clearly part of it has to do with personalities and the lack of adequate resources for someone to play an active role in guidance and leadership. The small-farm structure of the county, moreover, militates against widespread farmer involvement and interest. This is not to suggest that small farmers are necessarily antagonistic to conservation. Rather their time and inclination for attending meetings is less, and they are less likely to be convinced about the need for presenting themselves as conservation-minded (through FFWAG) to a sceptical public. Possibly the Berwyn incident did not help in securing an atmosphere favourable to FFWAG's development and certainly the organized farming groups in the county have not been particularly forthcoming in providing support. In the first four years of existence no invitations were received by Montgomery FFWAG for a representative to address a local NFU or FUW meeting. Nor have relations with other groups been encouraging. The Secretary reports two "antagonistic" meetings with Young Farmers' Clubs. Attempts to persuade agricultural teachers at a college of further education to use Conservation in Agricultural Education Guidance Group notes for their lectures have not proved easy either.

Another factor which should not be ignored is the relative success of the Montgomery Trust for Nature Conservation. A number of MSC-funded posts had allowed it to develop a certain amount of advisory work amongst farmers. This led, in turn, to the appointment of a Trust officer for farm advisory work supported by Countryside Commission monies which might otherwise have gone to FFWAG. The idea of seeking similar MSC support for a FFWAG officer was floated but this was resisted by national FWAG. This leads on to another reason for Montgomery FFWAG's weakness. It has been unable to convince national FWAG and the Farming and Wildlife Trust that its problems are special and that special remedies are needed. For example, a request to provide funds towards employing a part-time adviser were refused. Locally it is felt that national FWAG is somewhat inflexible in its ideas about the development of local groups based, as they are, on experience predominantly gained in the eastern and southern counties of England. As one FFWAG member pointed out during an interview, Montgomery, with an electoral population of only 40 000 and no major commerce or industry, has 1 per cent of its population belonging to the Naturalist Trust yielding an income of £2800 per annum. In Essex, with a considerably lower percentage of the population in membership, the Trust has an income of £70 000. Merging with other counties is not the answer as the area to be covered would be too large, and in any case neighbouring Radnorshire has an electoral population of just 17 000.

Notwithstanding all these external difficulties it is also evident that the group lacks a clear sense of direction and motivation. As one member put it in an interview "the basic obstacle is lack of will which springs from a lack of a clear idea of what the Group should do". Differences of opinion regarding the role of the group, have come to the fore in discussions on the feasibility of employing a full-time adviser. A number of members clearly felt that the county had no need of an adviser and that there would be little demand from farmers even if such a service were made available. Others were unhappy at the prospect of young and inexperienced outsiders being employed.

In the view of a few farmers interviewed, FFWAG was less a means of promoting conservation than of limiting the impact of conservationists on farming, as the following quotes reveal:

> Conservation to me is something that has got to be watched. It's becoming all-powerful, and getting to the stage where they dictate.

> I don't know if we should go as far as to encourage people to be conservation conscious. (Interviews, Spring 1984)

On the other hand, one member saw FWAG's role as developing into that of arbiter of land-use decisions and providing the means of implementing a national land-use strategy. The norm was somewhere between these two extremes, typically seeing FWAG as having a positive conservation role but definitely erring on the side of caution. As one member put it:

> In this area we must avoid being tarred with the conservation brush. FFWAG needs to become regarded as a central, level-headed organization, providing practical, sensible advice, not advocating AONBs and National Parks. It's not the sort of county where you can easily have a Conservation Officer. The Demonstration Farm at Castle Caerinion is a better example – it's well run with good stock, and it's pretty. If you get the idea across that way – you don't have to be a bad farmer to be a good conservationist. We have got time to play with here (i.e. in Montgomery). People are worrying about reseeding, but there's still a lot of time to play with. Provided there's no coercion and provided the local community is seen as the most important thing to conserve, most should flow from that. (Interview, Spring 1984)

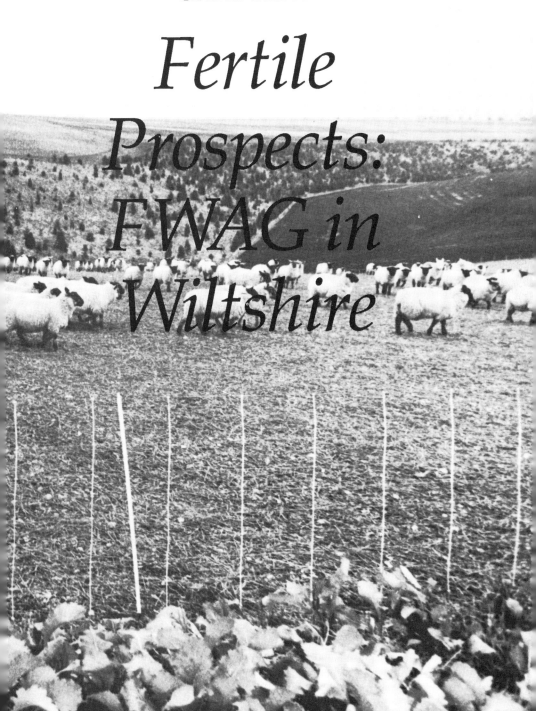

CHAPTER 7

Fertile Prospects: FWAG in Wiltshire

Introduction

On 10 April 1984 Bill Wilder, then Vice-Chairman of FWAG and Chairman of the Wiltshire Branch, presented evidence to the Sub-Committee on Agriculture and the Environment of the House of Lords European Communities Committee. Hansard records that after inviting Mr Wilder to say a few words in introduction its Chairman commented "You are the sort of peacemakers who are going to inherit the earth, are you?". "That", replied Wilder, "is a very nice idea, my Lord Chairman". But for critics of FWAG the noble Lord's inadvertent transposition of the New Testament message precisely captures the spirit of the sceptical reaction that the organization engenders. It is, of course, the meek who are going to inherit the earth.

Any advocate seeking to make an effective case for FWAG would be likely to point to Wiltshire. Certainly, since it became at the end of 1983 the first county to take advantage of the funding arrangements offered by the Countryside Commission to appoint an adviser, it has appeared to offer FWAG an image of its own preferred future, enjoying effective support, a stable committee and enthusiastic and charismatic leadership; indeed, its first chairman took the Silver Lapwing trophy as the winner of the *Country Life* Farming and Wildlife Award in 1980 and went on to succeed Dr Norman Moore as Chairman of national FWAG in 1985.

Indeed, if Montgomery – fragile, uncertain and beset by cruel fortune with the loss of its two key activists in its second year – seemed trapped in a vicious circle of ineffectiveness, Wiltshire quickly assumed a prominent trajectory. But to focus solely upon such obvious contrasts would be to miss many of the no less illuminating similarities. For, although Montgomery and Wiltshire present starkly contrasting cases, each illustrates in its particular way the common constraints experienced by the national organization as it attempted to insert FWAG into contexts characterized by long-established allegiances, inter-agency rivalry and competing initiatives.

Our concern in this chapter, however, is not only with the politics of establishing a county FWAG. Whereas Chapter 5 presented a snapshot of the network of county groups the emphasis here is more historical, showing the way, for instance, in which conservation is variously interpreted and negotiated within the emergent FWAG culture. Based upon extensive documentary materials, correspondence and committee minutes it also draws upon detailed semi-structured interviews conducted with the members of the Wiltshire group between 1984 and

1986. Their attitudes and understandings offer important insights into the nature of the FWAG enterprise and show clearly enough that whilst it does not represent a paradigm shift in thinking about the relationship between agriculture and conservation it can, for some, be part of a process whereby a more critical awareness is engendered. And just as it offers insights into the operation of a county FWAG before and after the appointment of an adviser, the case of Wiltshire illustrates too the promulgation of the distinctive voluntary ethic associated with FWAG. But it is inevitably, in many respects, a very particular story.

Establishing the County Group

Wiltshire covers 344 064 ha, almost half of it of AONB status. In the north-west soils tend to be heavy clay, but the lower two-thirds is chalk, and Salisbury Plain dominates the centre and south of the county. Much of the central part of the Plain is used for army training. Elsewhere on the chalk there are large arable and stock enterprises with dairy and stock rearing on the smaller farms of between 20 and 60 ha, which predominate in the north and west. In contrast, enterprises on the chalkland are often over 400 ha. Indeed, the estate at the heart of the Chalklands Exercise (see Chapter 3) was not untypical of its area of the county: an area often, with good reason, given the epithet 'Strattonshire'.

Roughly three-quarters of Britain's remaining unimproved chalk grassland is in Wiltshire and is of international importance as a wildlife habitat. It has suffered predictably from pressures to intensify production. Much of the ploughing began during and after the last war and, with the progressive application of artificial fertilizers, land which would normally have grazed sheep and cattle could be made to grow crops. Since much of the downland is on steep slopes, the further extension of arable cropping has often been effected by ploughing the bottom and top, leaving an unmanageable section in the middle which, ungrazed, reverts to scrub. The number of SSSIs in the county is slightly higher than average since there are no large upland blocks included. Under the 1949 Act 108 SSSIs, many of them downland sites, were notified and that number has now increased to over 130 making the proportion of the county notified in the region of 3.5 per cent. A study in 1977 showed that 35 per cent of the sites had been partially destroyed and 5 per cent totally obliterated (Tubbs, 1977). Since the war, according to the Wiltshire Trust for Nature Conservation's recruiting leaflet, the county has lost 50 per cent of its ancient woodlands, 80 per cent of the grasslands of the high chalk downs and 95 per cent of its flower-rich hay meadows.

The Wiltshire Trust was formed in 1962. By 1965, with over 2500 members, it was managing 35 reserves totalling nearly 360 ha, with an average size of about 10 ha, many of them being run on the basis of very informal gentleman's agreements. But whilst it owned 4 and leased 4 of its reserves it neither owned nor leased any chalk downland: an ownership situation which compares badly with other county trusts and seen by the then field officer (the third since 1974) as something of an indictment of its policies over many years (Peter Phillipson interview, 7 August 1985). In characterizing the Trust as "15 years behind in terms of its attitude" he referred particularly to the dominant influence of its Chairman, Michael Stratton, who had consistently been against the Trust buying and owning reserves and appeared to take the view that "only large landowners or farmers should have the right to hold land". Such acute sensibilities regarding the rights of private property accompanied by an equally stronge defence of the Trust's position in the county did much to stymie the initial attempt to establish a FWAG. Farmers and landowners had always been strongly represented in the Trust and along with, perhaps, a sense of satisfaction with its own activities there was certainly the feeling that the farmer involvement made another organization unnecessary.

In 1973 FWAG's Chalklands Exercise (see pages 27-31) had been held on the farms of Col. Jack Houghton Brown and David Stratton, nephew to Michael Stratton, which together extended to over 2200 ha and, given the impetus that such events gave to the formation of county groups elsewhere, it is somewhat surprising that a further four years elapsed before a county group was established in Wiltshire.

Following the Chalklands Exercise, Jim Hall sounded out the possibility of establishing a group in the county. Many of the thirty or so volunteers who had surveyed the area in the preceding twelve months were Wiltshire Trust members. But when Hall attended a meeting of the Trust's Council afterwards and tried to persuade them of the good sense of taking a more active part in wildlife conservation on farmland through a local FWAG, they showed little interest; indeed, one of their number left the meeting in disgust. There seemed, therefore, to be no space for FWAG in Wiltshire and in the face of this unambiguous rebuttal Hall did not pursue the matter. Later, the fact that David Stratton's farm at Kingston Deverill was on the short list for the Demonstration Farm Programme, which was drawn up by 1975, presented a constraint just at a time when – even allowing for his, no doubt, optimistic presentation of prospects – Hall seemed otherwise to have grounds for entertaining hopes. For whilst, by then, the attitude of the Trust appeared more accommodating the agreement with the Commission not to set up county groups where they might overlap with the project (see page 71) effectively foreclosed the possibility of taking too overt a lead in Wiltshire.

However, when Hall met the Trust's Field Officer, Peter Newbery, in 1977 for discussions regarding his tentative plans for a 1979 re-run of the Chalklands Exercise, Newbery indicated that the NFU County Secretary, Peter Riddick, had recently expressed some interest in liaison on the question of farming and wildlife conservation. Hall wrote to Newbery on 28 April 1977, one year after the first meeting of county committee secretaries (see page 71), indicating his anxiousness to pursue the idea, particularly since "My Group has now given priority to this Local Committee Movement". He could, he wrote, be "fairly certain of support from ADAS" and the proposal for a Chalklands re-run had met a favourable response from the other bodies which had helped with the original exercise.

Subsequent correspondence indicates clearly enough the delicacy of such local manoeuvrings, for Hall's carefully worded letter of 23 May to the NFU Secretary elicited a response which left him uncertain whence the initiative had come. According to Riddick he had been approached by Newbery who had advised that "the Wiltshire Trust thought there might be some advantage in closer links with the farming community". Then, at a later meeting with Newbery he himself had suggested that this might best be achieved by "creating a Wiltshire FWAG similar to that operating in Gloucestershire". He had further suggested that the Ministry be called on to initiate the inaugural meeting since he had "reason to think they might be sympathetic to this approach", adding that he would ensure the NFU was well represented (Riddick to Hall, 23 May 1977).

Hall, no doubt gratified that some move had been made from within the county, but by no means certain how best to proceed, contacted MAFF's Divisional Agricultural Officer in Gloucester who indicated that there would be no problem in ADAS initiating the meeting as they had done in Gloucestershire. Despite Riddick's account of the origins of the initiative an understandable wariness regarding the position of the Trust persisted. Hall adopted the strategy of channelling the request for participation of local naturalist organizations through the Wiltshire Natural History Forum, an umbrella organization set up in 1974 by Peter Newell who worked as Countryside Liaison Officer for the Community Council for Wiltshire.

Those invited to attend a meeting on 29 November 1977 included the County Planning Officer, the Principal of Lackham College of Agriculture, the NCC's Regional Officer, the Forestry Commission's South West Conservator, the ADAS Divisional Officer, the NFU and CLA County Secretaries, Miss Beatrice Gillam, who had co-ordinated the Chalklands Exercise survey work, and Michael Stratton. In enclosing a copy of the invitation letter Newell wrote to Hall on 20 October advising some prior stage management so that a possible Chairman and Secretary might be lined up, suggesting the names of Michael Stratton and Peter Newbery.

By the time he received that letter Hall had, however, already written to Lackham's Principal, Peter Walters, inviting him to chair the forthcoming meeting; reasoning that he would, by virtue of his position, have a valuable degree of independence. His 26 October reply to Newell's letter welcomed the positive interest being taken and took care to emphasize the importance of someone being prepared to take on and fund the secretaryship, for "the calibre of the secretary sets the tone of the committee". Newell had to report, however, that further discussions had shown a number of key people were "lukewarm to the proposal to form a local group" and had "substantial reservations about the efficacy of a local Committee". More groundwork in advance of the meeting was needed and he requested a more detailed breakdown of work being done by existing county committees, adding that another alternative being debated locally was that the Trust should set up a special sub-committee for the purpose (Newell to Hall, 8 November 1977).

Hall made no attempt to hide his disappointment at this news which seemed to herald a repeat of his earlier abortive attempt to elicit interest. In a long reply dated 11 November he pointed out not only that local committees could not, in the nature of things, begin with a clear idea of what they hoped to achieve – that would have to emerge from the development of mutual understanding. He emphasized, moreover, that such groupings were being accorded the highest priority by FWAG, active encouragement from the SPNR (the national umbrella body for the county nature conservation trusts) and, in the wake of the *Caring For The Countryside* statement of intent (see pages 39-40), they had never been more relevant. The Trust must be encouraged to approach the matter with an open mind. Hall copied Newell's letter and his own response to Wilf Dawson of the SPNR, speculating about whether his decision to channel the request for local participation through the Forum had been the right one, but asserting his determination to go ahead with the meeting and asking Dawson for any help that he might be able to give.

With the meeting drawing nearer other developments of note occurred. Neville Spink, ADAS Divisional Surveyor who had recently transferred to

Gloucester from Oxford, indicated his interest, thereby giving Hall the opportunity to comment in his reply that a number of committees had ADAS secretaries with the Ministry providing the secretariat. Then, following a discussion with David Rice, the Forestry Adviser in the County Planning Department, Peter Walters wrote saying that he and Rice had spoken to Bill Wilder whom they saw as a potential Chairman for the prospective group and who should, he suggested, be invited to the meeting. Hall's invitation was sent on 22 November and on the same day he wrote to Walters saying that the first meeting was very much "to sound the air" and that he had wanted to limit representation because he was fearing problems regarding the representation of the Wiltshire Trust. In the event Messrs. Stratton, Newbery and Newsam represented the Trust in the meeting at which 28 people were present, 12 of them from voluntary conservation organizations.

Wilder remembers the 29 November meeting at Lackham as "very stormy". In his introduction Hall recounted the origins of FWAG and emphasized the timeliness of the meeting given the recent reports from the statutory conservation agencies and the recently issued statement *Caring For The Countryside* which three members of national FWAG had helped prepare (but which, incidentally, makes no mention of FWAG). He explained the nature of local committees after which Michael Stratton, speaking on behalf of the Trust, said that agriculture was the Trust's main concern and he "wondered if such a committee was necessary since the Trust hoped to get funds from the NCC to provide advice to farmers" and others expressed fears about the prospect of "another talking shop" (Wilder interview, 11 October 1984). But in a sure indication that preparation for the meeting had been carefully orchestrated, the minutes record that after a positive discussion to which many contributed, and Hall's assurance that he had no wish to create a rival organization in the county, it was Michael Stratton who proposed, and Bill Wilder – the only farmer present – who seconded, the proposal that a local committee of FWAG be established. The proposition was approved unanimously.

A Chalklands Re-Run?

The process of convincing already established organizations that FWAG had both something distinctive to offer yet would not compromise territory they saw as their own, had been a painstaking one. But a consensus sufficient for the purpose in hand had been achieved. The composition of the committee was decided and the following organizations were asked to appoint members accordingly: six from the CLA/NFU to be mutually decided between them; one from the Young Farmers' Club; four from the professions – ADAS 2, Planning Authority 1, Lackham College 1; four members of conservation bodies – NCC 1, Wiltshire Trust 1 and Wiltshire Natural History Forum 2. Hall attended the first meeting on 8 March 1978 at which Wilder was unanimously elected Chairman. The offer by Neville Spink to undertake the work of secretary and arrange secretarial services was warmly welcomed. A subscription of £5 each from each organization would be asked for to provide working capital and David Stratton, like Wilder an NFU nominee, assumed the role of Treasurer.

With two others, Bill Isaac and Ken Fuller, ever present as well, the farmer membership of Wiltshire FWAG has not changed and that fact is significant given Hall's conception of the FWAG role. As Wilder put it, "We're lucky, our farming members are active: ours is a farmer core FWAG. Others are different. I suspect Gloucestershire is a MAFF core and others are conservation core". Moreover, whilst recounting his years in the Young Farmers, the "routine two-year shift" as NFU local branch Chairman and so on, he recollected that it had been "deemed prudent" that, as Chairman of Wiltshire FWAG, he become a Council member of the Wiltshire Trust (Interview, 11 October 1984).

Hall's immediate ambitions for the Wiltshire group were sounded at the first meeting. Wilder's recollection was that "Hall charged us with repeating the Chalkland Exercise five years on and I got the clear message that there was to be no Countryside Commission involvement". With the growing rivalry between FWAG and the Commission, however, this was not to be. Cobhams, the consultants for the Commission's Demonstration Farm Programme, had already approached Neville Spink and asked him to suggest a suitable chalk farm (Spink interview, 6 March 1985). He took Ralph Cobham, the project officer, to five farms. Despite it being rather large at 1160 ha – although typical, in that respect, of the farms in the area – Cobham was attracted to David Stratton's farm at Kingston Deverill precisely because of the extensive records which had been generated by the 1973 Exercise. Moreover, many of the recommendations emanating from the Exercise had proved a "dismal failure" so there was much scope for re-appraisal and future planning. A large committee was set up with Spink as its Chairman and since such developments were precisely contemporaneous with the renewed efforts to establish a county FWAG – and Spink certainly saw the developments as linked – pressing demands were obviously being made on local resources.

Hall's relations with the Countryside Commission were uneasy and the decision to make Manor Farm a Demonstration Farm was of immediate concern. Cobham, no doubt with the dual involvement of Stratton and Spink in mind, envisaged an opportunity to see how co-operation between the working party for the Demonstration Farm and Wiltshire FWAG could be arranged and, since his plans for the farm would not be completed by the time proposed for the re-run, he inquired whether it could be held without including Stratton's farm lest there be some confusion between the two. Hall, however, felt that comparability demanded that at least part of the farm be included and recorded in a memo for his Wiltshire file his feeling that "It is obvious that the Countryside Commission will go ahead with their project regardless of its effect on the Chalklands Exercise, and almost without regard for using resources which are already earmarked for the Chalklands Exercise: because of the Commission's superior pull we may suffer as a result". He had suggested that if the Commission was in earnest about co-operation with FWAG it should invite Wilder to attend a meeting between Cobham, Spink and Turner planned for the early new year (Memo., 11 December 1978).

Hall can have had little opportunity for further manoeuvre, however, because a letter from Spink dated 18 December informed him that the subcommittee, which the last Branch meeting had agreed should be set up to discuss arrangements for the Chalklands re-run, had met four days previously. It had begun with David Stratton announcing his agreement to include Manor Farm in the Countryside Commission project and "all those present were very pleased that this was being done". It was evident, moreover, that the meeting had shared Hall's qualms about whether there would be sufficient volunteers to carry out the survey work for a repeat of the Chalklands Exercise over the whole area proposed. It was thought preferable for work to be concentrated on Manor Farm with a view to making the information gathered available for a review "which might be held in 1981". A visit to Col. Houghton Brown's Lower Pertwood Farm was to be arranged for July 1979, meanwhile, so as to enable local farmers to look at the work carried out on the farm since the 1973 exercise.

Shortly after that visit, which he had been unable to attend, Hall expressed further anxieties about reconciling the promise made at the time of the 1973 Exercise with the new situation created by the Demonstration Farm Project and sought assurance that the survey work had not been forgotten. But his ambition for a Chalklands re-run which might get the new county group off to a fine start proved impossible to realize. The obvious difficulties proved insuperable. As Beatrice Gillam put it, "We all thought it impractical that the repeat could ever happen. Such a way away from everyone, so much work. We were lucky to have the people for the original exercise, half of them had disappeared . . ." (Interview, 1 March 1985).

Hall's promptings on another matter of crucial importance, however, met with a response which was to transform the capability of Wiltshire FWAG. In a letter to Spink, prompted partly by the minutes of Wiltshire FWAG's 11 October 1979 meeting which noted that David Rice, the County Forester, was getting more requests for advice than he could cope with, Hall wrote:

> FWAG nationally makes no secret of the fact that it would like to see a spread of people like John Hughes and Dorothea Nelson. It debated the issue at its last meeting, and I am sending you and the Chairman copies of the papers which were debated at that meeting. It may be premature for you to think immediately of any action, but there is no harm in thinking forward to the future and these papers may help. (Hall to Spink, 22 November 1979)

A Chairman and a Philosophy

Bill Wilder, Wiltshire FWAG's first Chairman, was interviewed some six months after he had given evidence to the House of Lords Select Committee. His first comment was that he would "love to know where FWAG will be in ten years" and it might have been supposed that someone so obviously given to "thinking forward" would also have direct experience of the organization stretching back over many years including, not least, the Chalklands Exercise. Such expectations would have been misplaced, however. Bill Wilder was wholly unaware of the Chalklands Exercise when it was taking place in 1973 and the process by which he became involved with FWAG was a very different one: but its most distinctive feature has an obvious relevance to present discussions about how the FWAG advisory effort should be developed.

Wilder's 160-ha farm lies on the Wiltshire-Gloucestershire border and its wide range of soil types from Cotswold brash to heavy clay lends itself to the 'half corn, half grass' cropping typical of the area. His grandfather had sold off the whole of what had been "a super sporting estate" in the 1930s. But his father managed to buy back the home farm and Bill Wilder, in turn, bought it from him during his last year at Harper Adams Agricultural College. Five years later, after three years working on the land in Australia and New Zealand, he took over with a "very hefty mortgage, so every square inch was farmed". Grandfather, however, had been a 'tree man' and much of the farm's 6 ha of mixed woodland, whose origins lay in covert planting, owed its continued existence to his enthusiasm.

Concerned to manage his high-input/high-output system as effectively as possible Bill Wilder recollects beginning to question whether he was managing the "non-productive remainder" of his land as "efficiently". "You become aware from the summer returns – the acreage on the meter on the drill – that there is a gap: the portion of your total acreage that you're not using. Here it was 10 ha. I felt I ought to do something with it". That sense of obligation came not from his college training which had in no sense prepared him for managing the natural features on his farm but, rather, from his wife's keen naturalist and ornithological interests. Though he met her whilst at college she was not, herself, from a farming family.

The effects of Dutch Elm disease made him conscious, too, of his need for advice on woodland management. At a 'Trees on Farmland' conference held at Stoneleigh he had met Geoffrey Laister, the ADAS Land Service man based at Gloucester, and was somewhat surprised when, after an interval of three or four months, Laister contacted him and offered to bring someone from the NCC and also a forester to look at his land. Laister, Philip Horton and David Rice from the County Council made their visit in 1974. Laister had not said a lot "but his presence gave credibility to the whole thing". Because he was there Wilder "didn't have to feel a cissy". The choice of word is significant and, as if to emphasize the point, he added that he "wouldn't have been seen dead around the place if they had turned up in an NCC wagon".

The three men walked areas which had been identified by Wilder beforehand and, as they did so, he became aware of connections between habitats and wildlife that he had not before perceived. But, crucially, they prepared a very detailed report for him identifying what was on the land and making recommendations as to its management and improvement. All he needed, as he put it, was "a time scale and to get on with it". He emphasized how tied to profitability the operations were. Returns from farming were good in the mid-1970s and that had made a lot of his conservation work possible. He had started to plant saplings in 1974 and the other easy things to get on with were pond clearing and coppicing. Spinneys which were bottomless from the treading of cattle were fenced off and a number of wet patches which he had been attempting to farm with great difficulty were planted with willows and allowed to reassume the character they had before his father filled them in.

Some 50 per cent of his woodland had been elm so there was a good deal of underplanting as well as new planting. Up to the end of 1981 he had spent £8000, recovered £5500 in grants and sold £3000 worth of trees. So his woodland operations had already by 1981 shown a modest profit. In the quiet and self-effacing manner which typifies his approach, however, he was quick to claim that he deserved little credit for the work which had gained him the Silver Lapwing Trophy since he had simply implemented the plan that had been provided for him.

What is remarkable about the story, of course, is the image of ADAS – later much maligned for its supposedly blinkered obsession with production to the exclusion of all else – actually initiating the preparation of a whole farm, conservation-oriented plan. If this practice had been even moderately prevalent within the county it would surely help account for that predisposition to question the necessity for establishing a Wiltshire branch. Our inquiries, however, quickly confirmed that the free and comprehensive service which Wilder had received was itself the product of circumstances that were both unusual and unrepeated. Asked if he had been involved with any other such plans David Rice found the suggestion almost fanciful and alluded to the expense – some £100 – of producing the report for Bill Wilder (Interview, 29 March 1985). Neville Spink, too, confirmed that the plan prepared for Wilder was a wholly atypical exercise. Indeed, Laister's situation was, itself, somewhat atypical. "Because of the staff that were in the Gloucester MAFF Office in the mid-1970s we could devote a senior surveyor to look after our land-use responsibilities and also conservation. Gloucester led the field in those days. In fairness most of it was land use but he was gaining expertise on conservation" (Interview, 6 March 1985).

The Ministry had gradually, Spink suggested, "been working towards doing more on conservation for years and years" but the nature of FWAG was very different because of the way a whole range of organizations were able to get round the table and talk about what they were doing. Hitherto, Ministry contact had invariably been with official bodies. Conversely, the Wiltshire Trust, which was active and had given some advice, "didn't have an official foot, as it were, into the farming system and were regarded with great suspicion by many farmers", who often resented claims about hedgerow removal which they regarded as misinformed. Around the Plain, for instance, there had never been any hedges because they would not grow properly on the chalk.

It is not difficult to understand the way in which Wilder's intense personal conviction about the necessity for FWAG was both nurtured and confirmed by his own experience and success. That conviction now expresses itself as a carefully considered philosophy which, particularly in its manifestation as a snappy slogan, was heard many times during the course of interviews with Wiltshire committee members. Wilder was asked whether, given that it necessarily saw compromise as a key virtue, FWAG could ever hope to have a clear line on matters relating to agriculture and conservation. Though, particularly in view of his recent evidence to the House of Lords' Committee, a ready and pat answer might have been expected his response was articulated only after very careful consideration. "Yes, it has to be the economics of the day. I must farm every square inch as hard as I can. Where it isn't being done that way you have to decide whether you're going to try to farm it efficiently and if not you should conserve it efficiently". Asked whether, in view of this, FWAG was only concerned with all the '10 hectares' left over he added that in relation to the rest there had to be "care, responsibility and, above all, moderation". Such notions, of course, sit uneasily with the injunction to farm as hard as possible and though they represent a rationalization of the fact that, as Wilder put it, "Our responsibility is to more than the 10 hectares, but 90 per cent of our advice will be related to those areas", they indicate, as well, the incipient contradictions of the voluntary philosophy.

As a way of eliciting interest in a hitherto unaware farmer the phrase "If you can't farm it efficiently, conserve it efficiently" has an obvious attractiveness. Instantly memorable, it invites a farmer to think about his farming operation in a new way as Alison Osborn, the County Adviser, confirmed. Conscious of the costs associated with conservation activity, particularly as effected through the negotiation of management agreements, she emphasized the need to get people to do things because they want to. "It's one of the things I'm pushing with the NFU. I mean, you can't farm all your land efficiently: and if you're going to farm it badly then conserve it well. That generally gets them thinking. They all know where these patches are. It's an easy thing to say" (Interview, 22 October 1984). She would be the first to concede, however, that as a conservation ethic it leaves much to be desired and the danger which understandably concerns critics of the FWAG approach is that this somewhat minimalist attitude becomes, by default, the most that a farmer might aspire to achieve on his land.

Clearly the hope of most conservationists is that somewhat different criteria from those which have dominated the post-war period might come to inform a more environmentally enlightened agriculture. The FWAG approach, however, can easily become a sort of veneer superimposed on an essentially unchanged farming philosophy. Indeed, as Peter Phillipson, the Wiltshire Trust's Field Officer put it, there is a tendency when you present the standard criticism of FWAG for farmers to "give it to you back with knobs on". Despite the presence in the county of some "pretty exceptional individuals" like Bill Wilder, David Stratton and others they still, he felt, "had that farming psychology. It's basically a farming psychology rather than a conservation psychology" (Interview, 7 August 1985).

During a period when circumstances have often contrived to polarize the view that farmers and conservationists have taken of each other there are good reasons why the FWAG philosophy has most often achieved a resonance with farmers who have already substantially completed programmes of farm improvement. Potter's study of FWAG members in Shropshire and Suffolk, for instance, showed many achieving a compromise between the competing demands of agriculture and conservation by segregating the two activities in time and space (Potter, 1987, p.155). Segregation in time means that conservation enhancement often takes place on farms with the most 'complete' investment histories and this pattern of activity was equally evident amongst Wiltshire members.

We need to recognize, however, the extent to which this view of the relationship between farming practice and conservation activity is seen by many farmers as a matter of moral obligation. The interview with Ken Fuller, one of the NFU's nominees on Wiltshire FWAG, for instance, began with him offering his memory of chairing the committee which had decided which county entry to forward for the 1980 *Country Life* Conservation Award which Bill Wilder won. He had, he recollected, wondered whether Bill and Jean Wilder had been over-zealous in their enthusiasm for conservation because they had actually taken out some 3.5-ton-per acre corn land. "Morally", as he put it, "you should farm your farm to the best of its ability, then look around it and blend in these odd areas, interesting areas and all those sorts of things". Fuller's farm, which is exclusively a dairying operation, runs to 62 hectares; now all owner-occupied after a period of renting from his father-in-law from whom he purchased the farm. He was at pains to point out the intensity of his farming and used words like 'aggressive' and 'severe' to characterize his approach. "By rights", as he put it, "there's no room for anything . . . and yet, and yet", he added with great emphasis, "we've got one and a half, probably 2, acres planted up, in corners, wet corners, odd bits . . . . " (Interview, 6 December 1984).

It is all too easy, of course, to dismiss FWAG as a 'trees in field corners' philosophy with the familiar litany of limitations which that implies. Indeed, at one point Fuller offered the example of taking out a hedge to make two 1.5 ha fields into one. Such a field is a better size for cattle, he explained, adding "then I clear my conscience by planting up a couple of corners and all of a sudden I've got more conservation in that 3 ha than I had by having a very severely cut hedge". As conservationists never tire of emphasizing, of course, it can be dangerously simplistic to suppose that newly planted trees and a long established hedge are of equivalent ecological value. But to focus only on such aspects of the conservation commitment of highly intensive farmers typifies in only a clichéd and one-dimensional manner the nature of the interest they have. Identifying the 'if you can't farm it efficiently conserve it efficiently' ethic simply indicates a predominant central tendency. A range of other factors can be significant in eliciting a conservation interest and the interview with Ken Fuller equally exemplified that point.

He talked with obvious delight about the natural interest on his farm and his anxiety to "get over to colleagues the tremendous satisfaction to be had". Education was obviously a major concern and his desire to get a mobile field classroom was just one of the many ideas that he presented during our long conversation. Educated at Lackham after starting his working life as an engineer with British Rail, he is now FWAG's representative on the Lackham Governors and does some extra-mural lecturing to third year apprentices. But more notable is his longstanding commitment to introducing young schoolchildren to life on the farm. On average some nine hundred 5 to 8 year-olds visit the farm each year, usually in groups of about 30. Part of his farm runs adjacent to a Trowbridge council estate and a belief that vandalism would be minimized if young people had an early introduction to the countryside provided some initial impetus for an activity which preceded the establishment of FWAG in the county by many years.

An impetus of another and very prevalent sort was evident on the 160 ha mixed farm owner-occupied by Bill Isaac, the fourth in the quartet of ever-present NFU representatives on Wiltshire FWAG. He and his three farming neighbours run a shoot and when asked whether his conservation interests pre-dated the establishment of FWAG in Wiltshire he spoke at length about the long history of tree planting on the farm. Love of the countryside was, he thought, the motivation since "none of us wants to see a prairie: we want to see a bit of habitat of some sort. And whether you are a hunting man or a shooting man you will put habitat there and look after what you've got . . ." Manor Farm has about 8 ha of woodland in all, including one 1.5 ha block planted for shooting in the late nineteenth century which was cut and replanted by Isaac's father and another 3 ha of 'old type' woodland which was cut very hard during the war and has since twice been substantially replanted with mixed hardwoods. Advice for all the forestry activities on the farm which involve 2000 new plantings annually has come from David Rice, the county Forestry Officer. He did not, as he had done for Bill Wilder, help prepare a management plan but, as Isaac put it, "We've walked the farm together so many times the plan is in my head" (Interview, 1 January 1985).

Trees and Grants

Whatever else they may say about conservation enhancement on their land – and there is much else besides – FWAG members are likely to draw attention to their tree planting and it is immediately apparent that the influence of David Rice in the county has been a considerable one. He is highly regarded by Countryside Commission staff and Richard Lloyd, their South Western Regional Officer, agreed with Keith Turner's assessment that few forestry officers in the UK have had the freedom and access to resources that he has had. David Rice found the characterization surprising: but he was able easily enough to account for it when he considered how much Countryside Commission grant aid had been spent within the county (Interview, 29 March 1985).

Richard Lloyd explained how he had himself been instrumental along with Lord Methuen in enabling Wiltshire to get the first agency agreement in the country for grant-aided tree planting (Interview, 19 July 1985). Following the Dutch Elm devastation at Corsham Court, Lord Methuen had written to *The Times* saying that there was no public support for the extensive replanting that was necessary. The Countryside Commission had, in turn, written to him pointing out that there was but that it depended on local authorities being prepared to contribute. Negotiations followed and eventually a scheme was evolved whereby the county would provide the manpower and the Commission the finance.

In this way an agency arrangement was instituted which was subsequently adopted throughout the country. Initially grants of 75 per cent were available in AONBs and areas where the elm had been a dominant species and by 1978/9, when funding problems caused an interruption, over 150 schemes were in operation in the county. Individual landowners put in the balance and Wiltshire, having a high proportion of large farms and estates offered plenty of scope. Indeed, the county spent more than any other, additionally introducing in 1976 its own Tree Planting Schemes in Towns and Villages. The pattern continued when the grant was reduced to 33 per cent and the figures for grant-aided tree planting in Wiltshire show that in the planting seasons covering the decade 1974-1984 some 783 schemes had been initiated resulting in the planting of 322 000 trees at an overall gross cost of £816 000.

Access to resources was only one element of Keith Turner's characterization however. He referred also to freedom and here the explanation offered by Rice contrasted the operations of the Forestry Commission and his own situation. Indeed, he saw his own position and the involvement of the Countryside Commission as testament to the fact that the Forestry Commission had left a particular niche unfilled. In the period before the initiation of the farm woodlands policy, they were not, as he put it, interested in anything under one hectare. And whereas they would now consider areas of 0.25 ha Rice, who took the view that they should be doing the job he was doing, felt that they had only made that concession for fear of losing out. The Forestry Commission, moreover, typically operated as he saw it with a civil service mentality: anxious to adhere precisely to the letter of their policies. His own situation, by contrast, was characterized by a flexibility in relation to policies: an ability, on occasion, to 'bend' the rules. Indeed, he recognized that the freedom of action he had enjoyed, in terms of discretion regarding what to do 'on the ground', had been "quite incredible" compared to the Forestry Commission with whom he had trained and worked for many years before taking up his appointment in Wiltshire in 1971.

David Rice emphasized that he had never had the time actively to encourage people to take up tree grants and the fact that he was, despite that, barely able to cope with his workload encouraged him that FWAG's strong commitment to a responsive philosophy need not prove a disadvantage. When the Group was established in the county there was, for a short period, a marked response and "since most people think tree planting is what they should do for conservation I had always to be involved". His experience then and since had convinced him that leaders in particular areas do latch on to ideas and then set a trend for their neighbours. There were areas in the county where a good few people near each other had planted trees and other areas where very little had happened. Ken Fuller, for instance, spoke of how, in his capacity as

Chairman of the Parish Council, he had initiated much grant-aided planting in his village and had spoken to adjacent parishes which had, in turn, planted such that "by us being a nucleus, all of a sudden it became a big thing" (Interview, 6 December 1984). Quite aside from such village-level initiatives there is considerable further potential for improvement in the county since there are some 2000 small woods under 10 ha in Wiltshire amounting to 23 per cent of its total woodland and many are not managed in any way.

There are good reasons, then, why there is a fairly widespread consciousness within the county of the need to plant and manage trees. Such activities are something that farmers feel that they can do, and the high probability of their getting some financial assistance provides an added attraction. The limitations of the prevalent tendency to see tree planting and conservation as more or less equivalent were, not surprisingly, recognized and highlighted to different degrees by the members of the FWAG committee. But inevitably, in the context of very limited resources an emphasis on trees reflected both the scope of what could be done and the important legacy of David Rice's work since the mid-1970s. As Beatrice Gillam recollected of the Chalklands Exercise, "It was aimed at farmers . . . . There was far too much forestry . . . that was the one positive thing that came out of it, that people ought to be planting trees". David Rice, she added, was probably the best organized of those involved (Interview, 1 March 1985).

Putting a 'Bod in the Field'

Responding to requests for advice on grant-aided tree planting presented David Rice with more work than he could, at times, hope to cope with. Indeed, a Wiltshire FWAG meeting on 11 October 1979 agreed that in view of the considerable difficulties he was experiencing in dealing with all requests for advice someone might be co-opted to give advice on forestry matters in the Kennet area. In fact, so far as advisory work specifically related to FWAG was concerned, David Rice spoke of an initial flurry of interest following the setting up of the county group followed by a relatively dormant period. The Group's advisory effort was handled very much in an *ad hoc* manner in that early period and there was substantial agreement amongst the members interviewed regarding its rather limited extent. Indeed, neither recollections nor assessments were discrepant even though members were having to rely on memory since no records of any kind had been kept. That was just one of a number of indications of the way in which Wiltshire FWAG had started by being, in one member's words, "a fairly loose organization" (Lesley Balfe interview, 21 March 1985).

At the first meeting of the newly formed branch Jim Hall, speaking of a committee's typical stages of development, emphasized that the definition of problems relating to wildlife, landscape conservation and modern farming should be the first priority for any county FWAG. The differences amongst conservationists themselves, and between conservationists and farmers and foresters would need to be ironed out if effective advice was to be given. Communication was the biggest problem, he suggested, with a need both to make farmers and landowners aware that advice was available and make them consider their role in the farming/conservation question. A request for advice having been received, there was a discussion at the meeting on 20 November 1978 on the principles to be adopted for responding to such inquiries. They would be forwarded to the Secretary who would then ask a local 'agriculturalist' and appropriate conservationist to visit and the Secretary was asked to draw up a list of names and areas for 'agriculturalists'.

The general recollection was that, on average, about half a dozen advisory visits had been made each year and that tallied with Bill Isaac's estimate that something between 30 and 40 visits had been made prior to the appointment of Alison Osborn as full time adviser in November 1983. He had, himself, made between 12 and 15: always in the company of one or more other members of the group. An effort was always made to "get a good mix" so that he might go along with David Rice, or Rice with Bill Wilder or, perhaps, Beatrice Gillam. CLA representative Charles Blackwood, with a 120-ha arable farm, had for instance made a couple of visits with the then Wiltshire Trust's Field Officer, Nigel McCarter, 'in charge' and saw himself as having been able to put a practical farmer's point of view. So that whereas the Field Officer would, if confronted by "a derelict barn in a couple of acres of nothing between two fields, say that nothing was wrong, I would say the area could be cleared out and suggest something that could be done" (Interview, 22 November 1985).

Clearly the team strategy was invariably adopted and Lesley Balfe, the Wiltshire Natural History Forum's representative on the Committee, emphasized that she would certainly not have made any visit on her own. Her assessment of the Wiltshire FWAG advisory effort in the early years was typical in being notably circumspect, a view shared by Beatrice Gillam whose judgement was that there "had not been very much impact" since "we were just about ticking over". Similarly, Julian Crane, the ADAS

Land and Water Services specialist who took over the Secretary's role from Neville Spink in 1981, felt that the group visits had not been very effective. They had not been able to give comprehensive advice and he thought it unlikely that any extensive work on a single farm had been done. Certainly, no overall management plans had been prepared.

Nigel McCarter outlined at a meeting held on 13 June 1980 a scheme for the preservation of wildlife in small areas of farmland whereby the owners would agree, as far as possible, to maintain their particular areas of interest and in return receive a commendation. This scheme proved attractive such that by the end of 1982, in the first four months of its operation, 15 areas were under consideration, though in the event only one such certificate was ever produced. In other respects the development of the branch assumed a well-tried pattern with resource constraints severely limiting the scope of what might be attempted.

Farm walks were organized and a sub-committee was convened towards the end of 1979 to prepare for a hedging and ditching demonstration. Articles were contributed to *Phoenix*, the county NFU journal; a winter meeting with the Wiltshire Trust was held each year at Lackham College and contacts were established with other organizations, such as the Forestry Commission and the Game Conservancy. These and other develpments were all very worthy so far as nurturing awareness and encouraging communication between farmers and conservationists were concerned, but ultimately their scope was limited given the advisory ambitions of FWAG. The years before the appointment of an adviser were, in the Chairman's words, "five years of moderate activity . . . we all despaired at times" (Interview, 11 October 1984). As Beatrice Gillam put it, "we couldn't do real work involving records without someone full time to do it". Progress demanded the unlocking of resources.

In December 1981, Nigel McCarter wrote to Richard Lloyd, the Countryside Commission's South-West Regional Officer, indicating that he was preparing a paper on the financing and administration of a Wiltshire Officer and requesting a copy of the Dartington study which was being prepared for the Commission (see pages 77-79). That report was not published or forwarded until nearly a year later when Richard Lloyd wrote saying he had been pleased to hear that the idea for an adviser in Wiltshire was "still alive". But in the meantime careful preparations were being made such that when Eric Carter, the National Adviser, attended the meeting of Wiltshire FWAG held on 16 September 1982 and stated that it was proposed to set up a charity, known as 'FWAG Ltd.' to assist with FWAG funding, it could be announced also that a meeting between the Finance Group and Wiltshire County Council had been planned.

Bill Wilder and Bill Isaac especially were concerned that the Council be involved, arguing that it should be prepared to put some resources behind the worthy sentiments contained in their Landscape subject plan. In the light of past experience, however, prospects hardly seemed auspicious. Derek Barber, for one, felt there was little chance of their pulling it off. It was, in Bill Wilder's words, "a year of hard graft". The first approach to the Council foundered, but it enabled a foot to be kept in the door and a second application – made after identifying the key decision makers and lobbying the Chairman of the Finance Committee who turned out to be a near neighbour of Wilder's – proved successful.

A critical impetus for the determination to appoint an adviser, however, had come from a conference held in Stockport which Wilder, Rice, McCarter and Crane attended at which Patrick Leonard, Assistant Director (Policy), had signalled the Countryside Commission's willingness to provide some sort of funding for countryside advisers. Such signals must, however, have been indicative rather than conclusive since the Commission had by no means finally resolved its position on such matters. Indeed, a lengthy internal memo following a regional officers meeting on 24 November highlighted the need for further thought on conservation advisers. It was thus into a situation characterized by a degree of uncertainty as to the future shape of policy that the application from Wiltshire FWAG for grant aid dated 1 December 1982 came along with a covering letter from the Chairman pointing out that Wiltshire was "still working its way towards putting a bod in the field".

In a memo to the Assistant Director (Policy) on 15 December Richard Lloyd, referring to a forthcoming meeting early in January 1983 to discuss further the question of grant aid for conservation work and advisers, asked to be informed as to what had been said at the Stockport conference since the expectation in the Wiltshire application seemed to be that 50 per cent of salary and travel costs would be provided for the first two years with a declining proportion thereafter. In commenting that the post was welcome and had been encouraged he added that he would wish landscape aspects to be more adequately covered in an expanded brief before any grant were offered.

As the successive draftings of a 'Job Description for Conservation Adviser' indicate, the need to respond to the Wiltshire application was an important element in the firming up of the new policy initiative within the Commission. Indeed, the decision in principle to support the application was made before the Commission's resolve to back the appointment of advisers on a large scale, and certainly before the financial details of such a policy had been finally worked out since it preceded the launch of the Farming and Wildlife Trust by over a year.

A letter to the Wiltshire Chairman on the last day of January indicated an 'in principle' decision to grant-aid the appointment, provided the Commission could be satisfied that its wider interests in conservation would be reflected in the activities of the adviser, at a likely level of 50 per cent for the first three years declining to 25 per cent by the sixth year. The attached job description, used verbatim in the application form prepared for the post, envisaged the adviser as the first point of contact for conservation advice within the landowning and farming community. In addition to providing wildlife and landscape advice the adviser should liaise with other sources of advice and direct enquiries for specialized advice to the appropriate person in the county.

Moreover, as well as assisting in arranging schemes of voluntary help for farmers and landowners, the adviser should assist in organizing visits to the Demonstration Farm and to any Link Farms which might be established. The adviser would, in addition, be expected to give talks to promote the objectives of FWAG and would be responsible to Lackham College for the provision there of an educational and interpretative service on farming and conservation. Finally, the adviser would be expected to report regularly to a Steering Group drawn from Wiltshire FWAG and to undertake any other duties that the Group might decide from time to time. Discussions concerning the post and the question of Link Farms, about which Bill Wilder had reported some apprehension lest they involve some restriction on freedom to develop, took place during the following months: a period during which the Commission was also negotiating with the newly established Farming and Wildlife Trust.

The revised 'Memorandum of Agreement between the Countryside Commission and the Farming and Wildlife Trust for a Farm Conservation Adviser (Countryside Adviser) appointment with the Wiltshire FWAG' which was issued in October 1983 specified that, in addition to the duties directly related to the appointment, the Adviser would be expected to contribute from time to time to the wider development of advisory services by attending seminars, writing reports on lessons learned and so forth. It was to be a demanding role.

But if the qualities required of the person to be appointed were nothing short of extensive the implications for the county FWAG itself were to be no less profound. Whatever else, the accountability demanded by the Memorandum of Agreement implied a more formal approach which, in turn, would demand elements of reorganization. Astute leadership and a degree of vision had put Wiltshire in the position of being the focus for the considerable attention now devoted to FWAG. No wonder that a draft plan for FWAG prepared very shortly after the appointment of Alison Osborn as County Adviser had, under the line drawn at the foot of it, a few lines titled 'Further Considerations'. One, under the heading 'Constitution', made a brief observation and posed a question: "At the moment we have no AGM, elections etc. Can we go on like this?'

The Shape of a New Era

The appointment of Alison Osborn, chosen from over 200 applicants, as the first full-time adviser under the Farming and Wildlife Trust Ltd., marked the beginning of a new era for Wiltshire FWAG. A farmer's daughter from Bedford, she had taken a degree in landscape ecology at Wye College and was, for 18 months previously, countryside officer for the National Federation of Young Farmers' Clubs at its Stoneleigh headquarters. A press launch and reception held on successive days early in December 1983 ensured a high profile for the appointment both locally and nationally. Reports typically presented the group as a buffer between extremists on both sides with the newly appointed Adviser emphasizing that whilst she had no 'stick', she did have a few 'carrots' available, such as grants from the Countryside Commission.

With a planned initial annual budget of £15 000 FWAG's main sponsors in Wiltshire, adding to the Countryside Commission's grant of £5500, were the County Council with £2500, the Wiltshire Trust with £500, the Royal Agricultural College, Cirencester, with £250 and local farm-related businesses which contributed £1000. On 6 March 1984, a meeting of the branch learned that following local fund-raising activities £1600 had been received and 20 covenants with a gross annual value of £2500 had been made with the sums coming from firms and farmers each being over £1000. But, even after taking into account grants and donations promised the group needed to find a further £4500 annually.

Such figures show clearly enough the transformation effected by the well-managed appointment of an adviser. But they illustrate no less clearly the scale of the fund raising task which confronted FWAGs. For even in a county where careful preparation had secured County Council support, where the Chairman as a Director of Wiltshire Radio had good media contacts, and where farmer committee members were well connected with relatively prosperous sections of Wiltshire agriculture, there remained, in the early months of the appointment, a gap still to be bridged. The hope was expressed at the 6 March meeting that four farm-based gatherings for small groups of farmers might each realise £250 in covenants, and the need to think of ways of involving land agents, contractors and agricultural suppliers was recognized.

The 'activities' and 'display' groups seen by the draft plan as necessary to the effective management of the annual programme were established at that meeting. Organized on a quarterly basis it encompassed a spring hedgerow demonstration and farm walk while a further farm walk during the summer would be accompanied by a specialist visit with a particular focus. Autumn would see further hedgerow demonstrations and training in tree planting and maintenance whilst the main event of the winter months would be the Lackham conference. Within a couple of months David Stratton was able to report to a Steering Group meeting that T.G. Jeary Ltd. had promised up to £500 towards the cost of a display package, and David Rice announced that a suitable site for what was to be a substantial hedgerow project had been located on a Wiltshire County Council holding at Dauntsey.

Ken Fuller agreed to become Press Officer at the March 6 meeting and an editor was found for the proposed quarterly newsletter – to be, somewhat modestly, no more than two sides of A4. It was at that meeting, as well, that serious thought began to be given to such issues as the membership and constitution of the county group, the election of its officers and the need to have accounts properly audited. A small group from the Steering Group would be appointed to consider these matters. In these ways certain organizational and operational initiatives, which might have been expected at earlier points in the six years preceding the appointment of an adviser, began at last to be taken. The particular sense of obligation accompanying the appointment of an adviser had provided critical stimulus. As Bill Wilder put it in a *Countryside Commission News* profile (January/February 1985): "We have got to secure long-term funds ideally through covenants over several years. After all we are not just thinking about conservation. We are dealing with someone's livelihood".

Operationally, a key decision, with equally significant strategic and resource implications, concerns the location of the adviser. Whilst the application to the Countryside Commission for funding to appoint an adviser had expressed a hope that the County Council might provide administrative facilities in the early days of the appointment, Wiltshire FWAG, as with so many other counties subsequently, based their adviser in the Ministry Offices. Responding to a query about the possibility of farmers receiving conflicting advice from two well-informed sources Bill

Wilder, in his evidence to the House of Lords' European Communities Committee, pointed to the advantages of the arrangement. The adviser "has only to walk out of her door to meet the drainage officer who perhaps has just been on a farm to discuss the drainage of a large acreage on which in one part difficulties may have arisen and the decision may have been not to drain, and he can say 'There you are; there is your next visit'. That relationship is working superbly".

That assessment, however much it might seem – given the example – to imply a purely secondary role for FWAG, was shared by Alison Osborn, who felt that she had won the confidence of the Ministry people such that they were pleased to be in a position, when anyone showed interest, to say "Ah, I know just the person you need to see" (Interview, 22 October 1984). Given the prevalent suspicion of local authorities – a characteristic of the farming community and FWAG which was of concern to the Countryside Commission – to have put the adviser into County Hall would, in Bill Wilder's view, "have been asking for trouble. It would have been quite the wrong emphasis whereas ADAS and the farmer are old friends".

Though she hoped to see FWAG become "the first port of call for the farmer so that we become the broker, helping to rationalize the advice that is available", Alison Osborn certainly found the Ministry crucial for referring contacts. Other means of generating interest were important – a talk she had given to Wiltshire CLA's AGM had led to 6 visits for instance – but, after the initial group contacts generated by FWAG members and FWAG activities, the greater part of her advisory work had come through close liaison with ADAS and with David Rice, who was based in the County Planning Department.

The initial stages of the appointment were inevitably much taken up with publicity and visits to members of the FWAG group. Two days were spent at Lackham to learn about conservation education at the College and to consider the areas of greatest potential on its estate for establishing the intended farming and wildlife trail. Similarly, visits were made to the Demonstration Farm for briefings on both the route around the farm and the adviser's future role in its operation. The frequency of advisory visits varied and the third adviser's report covering the period from 9 January to 23 February 1984 began by noting that although the winter months had been busy a lull had occurred at one point such that the Chairman had contacted some FWAG members to invite the adviser to their farms, only for the pace to pick up so quickly that all those visits had to be cancelled.

Though the Chairman and others were anxious that their advisor establish her own style, a question about numbers of visits when interviewed instantly prompted her to bemoan the fact that she was not going to reach the figure of 100 in her first year. That figure was fast becoming an informal norm following the activities of John Hughes in Gloucestershire and, whilst acknowledging that the quality of visits was a more relevant criterion, Alison Osborn clearly felt a pressure – no doubt partly self-generated – to reach that target. Instead, with the pace having slackened during the summer months she found herself, as the 12-month deadline approached, somewhat becalmed in the mid-90s. Of the 94 farms visited 39 had planted trees and a further 47 were intending to do so; 15 had dug new ponds at some point in the previous ten years and 14 had cleared existing ones; 16 were planning to dig new ponds in the next two years and 14 had old ponds which they were intending to renovate.

Commenting on such data as she had been able to produce Alison Osborn was understandably diffident and emphasized the uncertainties: "it's 'iffy . . . you can see a farmer and you might not think you've had a response, but you've set him thinking, then David Rice goes out there and says the same things, others say the same, he sees a TV programme, goes to a FWAG meeting and eventually . . . ." In other respects, too, there were gaps in the information which might be thought essential to the monitoring of FWAG's effectiveness in reaching a cross section of the farming community. MAFF protocols regarding confidentiality meant that the adviser, in reporting to her Steering Group for instance, could only speak of having advised a certain number of farmers and give the total hectarage which they represented. Clearly it was often possible to note whether a farm was owner-occupied or tenanted but information of a systematic kind regarding ownership type and farm size was not readily available. Moreover, without knowledge of the details of what she had said to whom, members of the Steering Group were not in a position to reinforce her advice in their own encounters with local farmers.

The pressure of responding to requests, moreover, meant that no consideration had been given to the question of how to make contact with particular categories of farmer or devising a strategy related to the structure of Wiltshire farming. The impression was that, whilst she had made visits to quite a few smallholdings and a good number of farms in the 160 to 200 ha and 280-plus ranges, she was short of people in the 30 to 80 ha range: "but I need to get a breakdown of what Wiltshire is like, I suppose. There probably aren't many in that size range". That impression was certainly correct and, in the immediate aftermath of the imposition of milk quotas, the relative absence of contacts in the lower range was hardly surprising.

Alison Osborn and the other advisers appointed in the early stages were in the position of having to define their roles as they went along. An early visit to Sandy for discussion with other advisers and Eric Carter's recognition of the need for a training programme provided some acknowledgement of this demanding feature of the job. Her fourth adviser's report for the period to the end of March 1984 began by noting that she was now finding it difficult to keep office work under control and still respond with reasonable promptness to farmers' requests for advice. Fourteen farms had been visited in the previous four weeks and along with the Demonstration Farm and Lackham commitments and a meeting with other FWAGs to plan a stand at the forthcoming CLA Game Fair, it was proving difficult to deal properly with the range of commitments. Moreover, an increasing number of visits were being made on a preliminary basis with a view eventually to producing an overall farm plan after further visits: but such return visits could not be contemplated for some months.

Though the question of charging for advice had been discussed *"ad infinitum"*, as David Rice put it, the Steering Group had always concluded that it would be a mistake – certainly unless everyone did – since farmers who were well used to receiving free advice might then simply avail themselves of the sources of free advice provided by commercial interests. The members of the Group did, however, see a pressing need to secure additional advisory assistance.

A document presenting the case for a trainee for Wiltshire FWAG to be funded principally by the Manpower Services Commission was prepared for presentation to the 1985 AGM. Its outline of what had been achieved was witness to the reverse causality which characterized FWAG adviser appointments. Whilst many county groups agonized over whether the level of demand for advice warranted the appointment of an adviser, evidence from those counties where one was in post indicated that with suitable support an appointment would quickly generate its own justification. In the first fourteen months of her contract Wiltshire's adviser had been invited on to 144 farms covering a total of more than 22 800 ha. On 37 of the farms the requests had been for whole farm management plans whilst the remainder had involved advice ranging from the care of a small pond to the felling and replanting of woodland. Detailed survey work by members of the Wiltshire Trust or by ADAS had often been undertaken and return visits by the adviser were involved in many of the cases.

In addition the adviser had spoken on 28 occasions to evening gatherings of farmers' discussion groups, NFU local branches, naturalists' groups and others. Moreover, as well as assisting at Lackham College by talking on several occasions to staff and students she had administered eight demonstrations, largely to farming audiences, at the Countryside Commission's Demonstration Farm and had generally promoted the work of FWAG through a number of press articles and radio and television appearances.

Shortcomings which warranted a further appointment, however, were evident. The virtual absence of a follow-up service for farmers who had taken advice but who had not decided to act upon it immediately was an unsatisfactory limitation which meant, in turn, that information about the take-up of advice was minimal. The ambition to make regular contributions to local papers and farming journals had not been realized and many areas of education and training, such as student projects at Lackham and Young Farmers' projects, remained untouched. Neither had any move been made to establish the Link Farms which had been suggested by the Countryside Commission at the time of Wiltshire's application for funds to appoint an adviser. The document also noted the absence in Wiltshire of any award for conservation on farms of the sort which existed in many other counties. Clearly Wiltshire FWAG had made considerable strides in increasing awareness of its aims since appointing an adviser: but aspects of her demanding job description remained only minimally developed.

Patterns of Accommodation

The paper presenting the case for a trainee envisaged that the office and clerical costs associated with the new appointment would be met by the Ministry of Agriculture. David Rice, it was decided, should present the case to the Manpower Services Commission which would meet the cost of wages through the Community Council, with the travelling costs being covered by the Farming and Wildlife Trust Ltd. The minutes of a Steering Group meeting held on 13 May 1985 record that support in principle had been obtained from the MSC and the Trust. A reply from MAFF in response to the letter requesting additional accommodation and secretarial assistance was awaited. The reply, when it came, was negative. Spink had been informed that the Ministry was not prepared to countenance the resource implications.

Indeed, as Philip Edmondson, the other MAFF representative on Wiltshire FWAG, correctly anticipated (Interview, 6 December 1985) cuts in funding were likely to mean that the Ministry's generosity towards

FWAG would have to be reconsidered: a situation he found embarrassing in view of the help which the Ministry had received, especially from David Stratton, and its dependence on the goodwill of farmers for sites for field trials and so forth. In the event David Stratton, in presenting the third annual report for the year ending 31 March 1986, drew attention to a further increase in projected expenditure accounted for substantially by the need to make provision for 'buying in' such things as typing and photocopying since the Ministry, following staff reductions, would no longer be able to provide all clerical support as in the past.

The substantial support for FWAG in Wiltshire and elsewhere, both through its provision of services in kind and the ready acceptance of secretarial duties by many of its officers, helped engender the presumption on the part of many a sceptical conservationists that FWAG is, in essence, a 'front' for the Ministry; an 'arm's length' way in which it could respond to the insistent demand that it embody environmental considerations in its work without having to effect too rapid and radical a shift in its basic position. Bill Wilder readily acknowledged the truth in that prevalent perception when it was presented to him. But he did not interpret it as a criticism. Rather, he justified it by claiming that whilst it was important for Ministry advisers to recognize both conservation potential and problems they should not themselves advise. He reasoned that, since most advice needed is fundamentally biological in character, it was appropriate for the Ministry to use FWAG as adviser because its own officers had, for the most part, an expertise in surveying.

Secretary Julian Crane, whose work as an ADAS officer was changing markedly, emphasized, however, that such surveying expertise – when applied to estate management rather than to the narrower preoccupations of farm management – had always embraced concerns with woodlands, hedgerows and landscaping. Indeed, in his view, it embraced all the elements now characterized as 'conservation'. Since being "encouraged", as he put it, to take an interest in conservation, he was in the process of being trained "to some degree" and had attended a number of internal courses aimed at helping ADAS officers recognize habitats and features which should be conserved. He was devoting a quarter of his time to the responsibilities associated with the new remit: a change made possible, given that staffing levels had actually fallen, by the fact that the capital grant work which had occupied such a significant proportion of his time had now dwindled to virtually nothing.

Though he had not yet, at the time of interview, accompanied the FWAG adviser on a farm visit he clearly saw ADAS as offering a very similar service which meant that he had to ensure that they worked together albeit, perhaps, with a different emphasis: a point that he elaborated by emphasizing his own very general interests as opposed to what he presumed would be the specialist biological and ecological concerns of the FWAG adviser. The training he had received was intended to enable him to recognize, for instance, a piece of downland worth preserving. He was not able, nor did he see it as necessary, to name the species growing on it.

Clearly his own work was in a period of transition and the element of associated uncertainty was evident. FWAG could not even, as he put it, have pretended to exist without the services which only the Ministry had been prepared to provide. Indeed, he was clearly somewhat resentful of the perpetual criticism of the Ministry and the way that FWAG members in general – though not in Wiltshire – seemed to be embarrassed to be associated with the Ministry: probably because they were acutely conscious of just how dependent upon it they were. He emphasized that the Wiltshire adviser had access to the full range of services, though not to files. MAFF had not, for instance, been prepared to provide information on hectarages visited except in gross terms: "when we say this is a confidential service, we mean it."

Crane had been conscious, when he first became involved with FWAG, of the need to be "wary" because of his limited knowledge. Indeed, he felt this had led to some misconceptions. "I can understand why they wanted to play us down a bit. Didn't want us to give them a bad image, if you will. But they, therefore, also didn't expect us to give them any technical help. That was the big problem. The image was that the poor old Ministry didn't have much to say and didn't know much and therefore they didn't include us: we provided all the services, but not much help. Of course, things have changed now and I believe we can give some assistance".

Stuart Lane, the NCC's Assistant Regional Officer, characterized ADAS as experiencing an "identity crisis": unsure of whether they should be helping county FWAGs or "getting on and doing the job themselves" (Interview, 14 August 1985). In Wiltshire the relationship has, nevertheless, been a notably co-operative one: an assessment evidenced by the fact that between 1985 and 1988 ADAS provided the greatest number of introductions for the FWAG adviser.

But those inclined to see FWAG as a much trumpeted, but ultimately hollow, success have rather more than the appropriateness of advisory divisions of labour in mind when they charge it with being a 'front' for the Ministry. Their concern is rather with the Ministry's involvement in processes of agenda setting and the way in which FWAG's commitment to a purely responsive and confidential advisory strategy seems at one and the same time both to mirror Ministry priorities and presuppose a piecemeal and incremental approach which cannot hope to be adequate in the face of the challenges confronting conservation. Such charges which, however perceptive, can have more than a whiff of paranoia about them are hard to substantiate. More relevant when considering this difficult question are the implications of the basic FWAG commitment to consensus politics. While one should be circumspect in generalizing from it, an episode concerning a proposed press release from Wiltshire FWAG does seem to exemplify the more subtle variant of the charge concerning the Ministry's influence.

The compulsory killing of badgers in the attempt to limit the spread of the cattle disease, bovine tuberculosis, met with fierce opposition from animal welfare groups and others who argued that the evidence for a strong link between the animals and the incidence of the disease could not justify the policy which the Ministry was pursuing. At the end of 1985 a review of the Government's policy by an independent team was placed in the hands of the Minister, though it was some time before it was made public the following year. Whilst it reaffirmed that badgers do constitute a potential wildlife reservoir of bovine tuberculosis it took the view that the disease is widely, but sparsely and unevenly, distributed throughout badger populations rather than contained within discrete "pockets". The report noted the many gaps in the epidemiology of bovine TB in badgers and, with no systematic evidence to confirm the transfer of the disease from badgers to cattle, it therefore considered the complete and permanent eradication of TB in badgers, and hence in cattle, unattainable.

In view of this conclusion the Review's recommendation, that a reduced programme of control related only to herd breakdown be undertaken, itself came under sharp criticism, not least from the NCC which questioned how an unsuccessful policy could be made successful by carrying it out to a lesser extent. In its view the only appropriate response to the evidence would be to stop killing badgers. Thus the issue had been, and remains, a highly contentious one with a substantial body of expert opinion ranged against the Ministry.

Eunice Overend, the county trust representative on Wiltshire FWAG, had been closely involved with the welfare of badgers since 1967 when she had begun hand-rearing and releasing them to combat the impact of badger digging. She had also worked tirelessly to convince the Ministry that, in the light of the available evidence, its suppositions about badgers and bovine tuberculosis were mistaken.

Having been considered at the Steering Group meeting held on 18 December 1984, an item was added to the fourth Wiltshire FWAG Newsletter circulated later that month drawing attention to a report prepared by the Badger Working Group of Wildlife Link for submission to the Minister's Review Panel. The Newsletter referred to it as a comprehensive and moderately written document whose recommendations might "form the basis of a thorough rethink of our whole approach to the problem". The steering group resolved that Eunice Overend should write an article for the next issue which would have to be approved by other members of the group, particularly MAFF, and that she would liaise with the Chairman to produce a draft press release which could be considered at the next Steering Group meeting before being finally approved by the full FWAG.

The Steering Group meeting on 5 February 1985 determined that the Press Release should be sent to all members of FWAG and it was included with circulated documents accompanying the agenda for the forthcoming AGM. An attached explanatory note from the Secretary, Julian Crane, listed the documents and indicated that any comments regarding the release should be sent to the Chairman by 4 March, adding that "After that date it will be assumed that members who have not contacted Bill Wilder agree with the contents of the proposed Press Release".

Considerable care and effort had been put into the drafting and redrafting of the Press Release on what was acknowledged to be a sensitive issue. Running to 388 words its closing paragraph began with the observation that "One of the problems of the present policy is the perpetual antagonism that it causes between farming and conservation interests", and continued, "The Wiltshire Branch of the Farming and Wildlife Advisory Group, which is a coming together of those interests in the county, has discussed this report and concludes that it shows a very balanced and moderate approach to this complex problem . . . "

The minutes of the 1985 AGM simply state that Neville Spink objected on behalf of ADAS to the proposed Press Release and that, after discussion, it was agreed not to issue it on behalf of Wiltshire FWAG. In fact, although the Press Release contained only one reference to the Ministry in the context of a straightforward statement of the then policy, Spink claimed not to have seen the release and he indicated that, following enquiries which Julian Crane had made of Ministry veterinary officers, he wished to add a good deal more material to it. Bill Isaac, one of the NFU's nominees, then proposed that it should be sent out with a note saying that it did not necessarily represent the opinion of all members of the FWAG committee. But Spink and Crane again objected and it was decided that there was little point in continuing. To do so, it was claimed, entailed the danger that anyone might think they could publish anything: a situation that would be sure to engender disagreement.

Wiltshire FWAG's first attempt to issue a collective statement on an issue had thus foundered on the objection of a single member. Moreover, since the opportunity to make representations before the date specified had not been taken it is hard not to interpret Spink's intervention as an exercise of veto power made possible by the commitment to consensus politics and designed to ensure that no statement which might be construed as critical of the Ministry could emanate from Wiltshire FWAG.

A commitment to consensus politics is easier to sustain in the absence of major issues of conflict, and a number of committee members, when interviewed, emphasized the relative calm of Wiltshire when compared with the fundamental confrontations that had, for instance, emerged on the Somerset Levels. On a committee which, in terms of relative input, could be characterized as farmer-oriented, the general agreement with the key features of a non-interventionist, gradualist, consensus-oriented, voluntaristic philosophy was striking. David Rice did, however, set his comment that "in the main if you want to make progress it has got to be done voluntarily" in the context of the framework established by the *Wildlife and Countryside Act* 1981 and the presumption that the existing structure of landownership was not to be questioned. In characterizing the conservationists on the committee as "mild" he emphasized that, as overworked amateurs, they had little opportunity to make more than a minimal contribution. Moreover, the "professionals" – the Trust and especially the NCC, who were totally absorbed in the mammoth task of renotification demanded by the 1981 Act – were too preoccupied with their own concerns to be able to do much.

A degree of scepticism regarding the possible effectiveness of FWAG was certainly evident on the part of the committee's representatives from the voluntary conservation organizations. But it was never accompanied by a thorough critique of the voluntary philosophy: nor were such limitations as might be hinted at linked to any systematic structural explanation of the factors setting agricultural and conservation interests at odds with each other. None, however, had been happy with the *Wildlife and Countryside Act* and Beatrice Gillam, Eunice Overend and Lesley Balfe all expressed concern as to whether the NCC would be given the funding to fulfil its statutory obligations.

It would be quite wrong, therefore, to present them as wholly uncritical: what was evident, rather, was the very particular and parochial nature of their concerns. The relationship between those concerns and broad policy developments did not really engage their attention. Indeed, as Peter Morris, successor to Peter Walters as Principal of Lackham College, recollected, the committee had never, for instance, really discussed the pros and cons of the Wildlife and Countryside Bill. They had been much more concerned at the time with, as he put it, "the nuts and bolts of getting an officer" and, he added, "That seems rather peculiar now, on reflection" (Interview, 27 March 1985).

The farming members of Wiltshire FWAG were notably more prepared to elaborate on the legislative and policy context within which FWAG was attempting to make an impact. The reluctance of farmers to acquiesce in controls of any kind was seen as a constraint on the policies which might be effective and such considerations underscored a strong commitment to making progress by persuasion and example. Indeed, FWAG's role in promulgating the voluntary philosophy amongst leaders within the farming community encourages a strong belief in education as the only appropriate way forward. As Bill Wilder put it, "The whole argument of the *Wildlife and Countryside Act*, planning controls, SSSIs and so on boils down to the need to influence the man in control of the land, so he wants and will enjoy looking after it for the sake of wildlife. Then you don't need controls".

If the evident discrepancy between the immediate nature of problems of habitat destruction and so forth, and the essentially long-term influence of education in modifying behaviour was pointed out, the response was usually to assuage doubt by reasserting, however hesitantly given the complexities of the issues, the core tenets of the FWAG philosophy. Again, Bill Wilder was typical in likening the process to the drip of water on a stone. "I think I disagree with you . . . it has to be trust . . . I think, you're wrong . . . you've got to want to take advice". As Peter Morris, put it, "It is a better way of approaching problems than any alternative. The philosophy of the 1981 Act is voluntary and FWAG corresponds to that perfectly: given the conservatism of farmers it is basically the right strategy".

The application for funds to appoint an Adviser had, of course, initially envisaged the post being combined with one at Lackham College so that the interest in landscape and wildlife management there might be developed. But the element of Alison Osborn's job description which made her responsible for providing an educational and interpretative service on farming and conservation, was one which remained relatively undeveloped in the early stages of the appointment. So far as actual involvement with teaching was concerned she had made it clear at interview that she was reluctant, given the sort of hostile reception a conservation lecturer had received when she was a student. Nevertheless the one talk which she had given to students taking the National Certificate in Agriculture had been well received and she had spoken to all course tutors about the need to include conservation considerations in all courses and to make use of the leaflets prepared by the Conservation in Agricultural Education Guidance Group (CAEGG, see page 48).

Clearly, however, given pressures on time and the large body of information that often needs to be imparted during a course there is always the danger that such considerations will be marginalized to the point of invisibility. Conservation can amount to little more than the emphasis on the need to be conscious of safety when dealing with chemicals. Peter Morris had, in David Rice's view, effected a significant shift of emphasis at the College and the new expansionist and more commercially oriented approach which he encouraged was much approved by the local farming community. In the context of such changes, of course, the issue of conservation teaching in agricultural college courses raises questions which go beyond the involvement or otherwise of FWAG and its advisers which concerns us here.

Lackham's Principal had come to the college in 1978 from the East Riding of Yorkshire where he was Vice Principal at Bishop Burton College with Graham Suggett, the CAEGG Chairman. He embarked on our discussion with the question "Are you a Shoard, Melchett, Body type?" and proved to be the most hostile of all the interviewees. Although he asserted in blanket fashion his agreement with all aspects of the line pursued by the agricultural lobby in relation to such matters as compensation and planning controls he agreed to make a formal "statement" about conservation, adding "You'll get a different line from Ted Culling": a reference to the other member of staff on the FWAG Committee and a dissenting voice on the College's board which decides academic policy.

His statement, which went on to outline at some length the role of the leaflets which had been produced in association with FWAG, began by clarifying the academic board's view of the place of conservation teaching within agricultural courses. That view, entirely consistent with FWAG and CAEGG policy, was that it should be "integrated within the husbandries" so that relevant aspects of conservation could be mentioned at appropriate points rather than there being something special in the curriculum called ecology or conservation.

He talked at length, too, about the way in which they endeavour to "practise what we preach" on the college estate. Recognizing that they enjoyed the benefits of its having been a traditional country estate in a beautiful location he outlined the ten-year Landscape Plan, "what I suppose we would now call a landscape and conservation plan", which had been drawn up in association with the County Planning Officer in 1976 and the new forward plan which had just been initiated. Both plans involved, in particular, hedge and tree planting – especially in field corners – as well as a considerable amount of new landscaping: there was, however, no mention of the squeeze on college woodlands which David Rice had alluded to and which was a consequence of some of the expansion plans which were in train.

Ted Culling, who characterized himself as "a strange individual" since he was a trained agriculturalist who was also a member of a natural history organization in the county, had been teaching an ecology and conservation course for three years but one of the external assessors had raised questions about the time devoted to such issues. In fact, relatively little time was devoted to the course. But in the subsequent discussions prompted by the question the integrationist view had, for entirely pedagogic reasons, prevailed.

Culling found himself wholly sceptical about the prospects. It was, he felt, impossible in the context of courses aimed at teaching students to farm with maximum efficiency to introduce other considerations which implied a very different set of instructions. As a lecturer on Grass and Management himself he was keenly aware that maximizing production would not, for instance, be in the interest of meadowland or other habitats. But he could not see how you could teach effectively telling students that x grammes of nitrogen could be applied to grass and then qualifying it by saying "Ah, but you might do y or z". Whilst recognizing the dangers of compartmentalizing subjects he felt a distinct course was the best way to create awareness of an issue so that people could then make their own decisions according to the balance they chose to strike.

The dilemma is a very real one. The case for integrating conservation into normal lecture courses recognizes the danger that agriculture students will dismiss conservation and ecology as irrelevant unless it is presented as part of their main subject area. Moreover, its credibility is enhanced if it is considered by regular college lecturers rather than being the subject of talks by outside speakers. Again, there is a danger that unless integrated with main agricultural topics, conservation will be seen as a limited cosmetic exercise involving, for instance, little more than tree planting.

Nonetheless, a study by John Beynon in 1983 (Beynon, 1984) of conservation teaching on National Certificate in Agriculture courses found that only a third of the 29 colleges included in the study who expressed a view thought that integration of conservation into all subjects was the best way to achieve a conservation input. Some 30 per cent thought that inviting specialist visiting speakers was the most realistic approach whilst 24 per cent thought farm visits a more appropriate method. Given the intensive nature of teaching on many courses and the fact that 45 per cent of colleges identified insufficient time as the reason why conservation had not already been included in the NCA course Ted Culling's concerns were well founded. The commitment and level of knowledge of staff matters a great deal as well, and again there were good grounds for questioning whether ecology and conservation would be treated in any but the most trivial and superficial manner.

The CAEGG leaflets are intended to raise levels of both awareness and knowledge and they are written so as to enable agricultural college staff easily to include them in their specialist lecture notes. But given the many constraints identified by the Colleges, the desirability of specialist teaching of ecological and conservation issues is clear enough. Cumulative experiences, moreover, are more likely to have a sustained impact on consciousness and understanding, and contributions from the FWAG adviser and effective use of the Demonstration Farm at Kingston Deverill are a necessary part of that process.

FWAG is, above all, concerned with the processes of accommodation and the ways of achieving such change are by no means mutually exclusive. The dilemmas inherent in such processes are apparent in the characteristic FWAG consciousness. Ted Culling, who was anxious to emphasize that ecology should be taught because there is a great deal more to it than just conservation, was equally prepared to characterize his own approach in terms of the distinctly FWAG philosophy whilst acknowledging, from a naturalist's point of view, its severe limitations.

He encouraged students, he said, to have the typical farming aspiration to see an animal or crop grow as well as it might. Farmers look for perfection, but they should also look to the well-being of everything on their farm. With that idea established the students could then be told "Now try to adopt a different stance and think about something in a different way and think about the balance of nature, the habitat. Instead of a perfectly growing monoculture of cereals you've now got a pond or a ditch, or some other sort of habitat, where there is a different form of perfection; a perfect balance, or something approximating to it. And I think it is perfectly feasible to adopt these two views and, as an educationalist, try to help the pupil see the possibilities and choose for himself. And that is how farmers can have highly farmed land and at the same time be looking after the environment. That is what FWAG preaches".

Consolidation and Advance

In his first AGM report as Chairman in 1985 David Stratton was able to highlight the progress which had been made since Alison Osborn's appointment. Over 150 farms had been visited in eighteen months and, in addition to the follow-up visits that had been made to develop the conservation advice that had been given, there was a welcome higher percentage of farms looking for an overall management plan rather than seeking to concentrate on small areas or specific projects. Indeed, that proportion which had risen from 22 per cent in 1983/4 to 55 per cent in 1984/5 further increased the following year to 62 per cent. Whilst welcome, the trend had considerable implications for the Adviser's workload: a fact emphasized by the sharp rise in return visits from 16 in 1984/5 to 51 in 1985/6, both years in which specialist advice from other countryside agencies and voluntary bodies was called in on 18 occasions.

By 1986 Alison Osborn had visited farms totalling over 31 000 ha, some 10 per cent of the farmed area of the county. Effective communication had been established with the main bodies involved with agriculture, land management and conservation and the Adviser noted in her report the enhanced co-operation with land agents and agricultural and forestry contractors as well as Wiltshire FWAG's good relationship with the Agricultural Training Board. That body was, for instance, the source of initial contact with 5 farmers in 1986, a year in which ADAS provided some 17 contacts: more than double the number provided by FWAG's next most significant source.

Of particular note was the progress that had been made with the setting up of a hedgerow demonstration site at Union Farm, Dauntsey. FWAG is particularly associated with hedge management, as well as with tree planting and pond management, and together with woodland management, these interests accounted for no less than 69 per cent of all requests for advice in the four years from 1985 to 1988. Although FWAG otherwise had managed only the most minimal involvement with tenant farmers, what was evident from many of the interviews was the obvious excitement at the prospects afforded by this County Council

Table 7.1 Sources by which initial contact between Wiltshire FWAG and farmers was made

	1985	1986	1987	1988	Totals
FWAG members	14	8	8	9	39
FWAG Activities	5	5	17	14	41
Personal Recommendation	10	5	13	13	41
Shows, Demonstrations, Talks	3	3	4	1	11
ADAS	14	17	13	8	52
Wiltshire County Council	10	7	6	11	34
Press	8	4	4	10	26
NFU	5	-	2	4	11
CLA	3	1	2	2	8
Countryside Commission	1	-	1	1	3
Contractors	1	3	5	1	10
Agricultural Training Board	1	5	2	1	9
WTNC	-	3	5	4	12
Land Agents	-	3	4	2	9
NCC	-	1	1	2	4
Kennet District Council	-	1	1	1	3
Wessex Water	-	-	1	-	1
Cotswold Park Warden	-	-	1	-	1
BTCV	-	-	-	1	1
Unknown	16	7	-	-	23
Total	91	73	90	85	339

smallholding, where hedges had simply not been touched for twenty and more years. Three successful demonstrations had already been held on the Earl of Shelbourne's Forest Gate Farm and the new site was seen as offering great potential for developing a very important aspect of the group's activities.

The County Trust's Field Officer sounded a note of caution, however, recollecting how the early negotiations had indicated both a somewhat blinkered mentality and – what was for him – a disturbing readiness to compromise where conservation values of the highest order were involved. Initial inspection of the site had revealed, of course, its spectacular suitability for the hedgerow exercises envisaged. But when one of the Field Officer's survey leaders was called in to do a detailed survey of the hedges, she found that the meadows on the farm, which had not even been mentioned, were of the unimproved ridge and furrow type and supporting the green winged orchid and other species now very rare in Wiltshire. To focus on a secondary habitat and not notice something

Table 7.2 Subject of requests for advice from Wiltshire FWAG, 1985-1988

	As Part of Whole-Farm Conservation	As Specific Subject	Total
Tree Planting	102	84	186
Woodland Management	54	34	88
Pond Management	67	62	129
Pond Creation	22	26	48
Wetland	24	3	27
Grassland	38	7	45
Scrub	8	1	9
Hedge Planting	12	8	20
Hedge Management	60	11	71
Water Courses	5	1	6
Pollarding	5	1	6
Green Lanes	1	1	2
Walls	4	1	5
Ancient Monuments	9	2	11
Wild Flowers	11	4	15
Organic Farming	4	-	4
Public Access	2	3	5
Others	10	3	13

unique and irreplaceable seemed, to the Field Officer, indicative of FWAG's prevalent concerns and their attendant limitations. And the initial suggestion from the FWAG Adviser that perhaps 25 per cent of the meadows be retained seemed a wholly inappropriate capitulation when other avenues such as rent reductions and compensation were there to be explored.

Part of the problem, of course, and one also emphasized in a number of interviews, is the relatively under-surveyed nature of Wiltshire. There had never, for instance, been a basic, Phase 1 survey of the type that identifies areas that might be of interest. Work to manage and conserve habitats depends, in the first instance, on knowing where they are and although some work of that type was being carried out by the Trust, with the support of the Manpower Services Commission, much remained to be done. Indeed, the FWAG Adviser saw as a major part of her role the need to encourage farmers to identify and become aware of what was of conservation value on their farms. FWAG's development and advisory effort has necessarily been a learning process for all involved: not least for the Adviser who is obliged to encompass a broad range of competences and areas of expertise, both technical and legal: a fact reflected in the increase to 15 of in-service days of training during 1987-1988.

The Hedgerow Management Demonstration Project which involved a considerable amount of the Adviser's time over a number of years proved very effective with over 200 people attending an open day held in November 1985 and some 130 benefiting from the second open day held a year later. The detailed monitoring of the effects of the management on wildlife by a FWAG team yielded valuable information and the County Council's Planning Committee took measures to ensure the protection of the flower-rich meadowland on the site. Joining forces with the Agricultural Training Board, Wiltshire FWAG was able in 1988 to run the first hedge-laying competition in the county for many years and it is now an annual event.

Interest in pond creation and management which was so evident in the county was sufficient to warrant the setting up, with the support of FWAG and MAFF, of a Wiltshire Ponds and Lakes Association whose inaugural meeting attracted over 120 people and whose 50 members showed a high commitment to the meetings and visits subsequently arranged. In other ways, too, the county FWAG did much to initiate events as well as take part in those arranged by other organizations. Indeed, with specialist advice regularly called in by the Adviser and the setting up in 1987 of regular discussions between FWAG, the NFU, MAFF, the NCC and the Wiltshire Trust for Nature Conservation, the Wiltshire FWAG more nearly operated in the manner envisaged by the Countryside Commission as first point of contact for farmers seeking conservation advice than did many other county groups.

Table 7.3 Responses to advice from Wiltshire FWAG, 1985-1988

	Number of conservation decisions taken where		Return visits	
	sporting interest was a factor	availability of grant aid was a factor	return visits made	requests outstanding
1985	19	31	16	13
1986	7	28	51	14
1987	15	52	38	5
1988	9	44	32	13

Moreover, its relations with the County Council – something the Commission was also anxious to encourage – were particularly strong: reflecting not only the care which had been taken to ensure its support when the group had been formed but also the pivotal role played throughout by David Rice. The County Council's commitment to the continued role of FWAG in the county was manifested most significantly by its undertaking, which was announced by the Chairman at the 1988 AGM, to fund 50 per cent of the Adviser's salary and travel costs when the Countryside Commision grant ended in 1989. Funding is, inevitably, a matter of continuing concern to FWAG; especially when an effective county group seeks to increase its range of activities.

Wiltshire FWAG's budget for 1986 to 1987, for instance, showed a rise in expenditure of about £2500 over the previous year, which was accounted for largely by spending on the hedgerow demonstration site, and the Chairman was forced to note in his report the following year that hoped-for sponsorship for the site was still not forthcoming. With the Countryside Commission grant reducing from 50 per cent to 40 per cent from November 1986, the fact that a high proportion of the original four-year covenants towards employing the adviser were renewed was particularly welcome, therefore, and the group continued actively to approach those who had received advice in the past. The funds thus raised by Wiltshire FWAG in 1987/88 stood at £4401 against the figure of £5745 which had been budgeted, and at a time of acute pressure on farm incomes, the forward commitment of the County Council gave it a far more favourable placing than the generality of county groups who faced considerable uncertainty.

Transformed by the appointment of an adviser and the extensive organizational changes which that development necessarily implied, Wiltshire FWAG had, by 1987, acquired a high profile within the farming and landowning community in the county. It was, in its Chairman's words, "a very small organization" which had "accomplished a great amount in the last few years", not least by presenting "a coherent and united story to the public at large". The attendance of the Adviser at various events each year and her very full programme of addresses to a range of meetings, in additon to showing groups totalling 800 people in three years around the Demonstration Farm at Kingston Deverill, ensured the wider promulgation of the ethic of compromise which FWAG seeks to embody.

Such were the demands placed on both the Adviser and the other members of the committee that it was not, perhaps, surprising that they had nothing to say when invited in interview to indicate what they would like to see Wiltshire FWAG do that it was not doing already. Just as there was little sense of any critical questioning of the adequacy of a purely responsive strategy dependent on voluntary co-operation, or often,

Table 7.4 Management type and hectarage of farms visited by Wiltshire FWAG 1985-1988

	Owner occupier	Tenant	Manager	Total no. of farms	Total hectarage
1985	60	23	8	91	11866
1986	51	10	12	73	10327
1987	68	12	10	90	9520
1988	64	9	12	85	13166
Totals	243	54	42	339	44879

indeed, of the history of FWAG – neither, it seemed, was there the time or the inclination to engage in forward thinking about FWAG's activities and role in a rapidly changing policy context.

An emergent critical awareness was most apparent when farming committee members were invited during interview to comment on Barry Wookey, a prominent Wiltshire landowner who began to convert his 660-ha farm at Rushall to organic production in 1970 (Cox 1981, Wookey 1987). The scale of the enterprise, the duration of the commitment to organic production and Wookey's ceaseless lobbying meant that no-one was unaware of his activities. Bill Wilder, for instance, said that he would "dearly love to be able to do what Wookey is doing because what he is doing had to be right for the land in every sense of the word. But few have the knowledge to farm properly now". The loss of traditional farming skills in particular made it quite inappropriate for FWAG to encourage people to farm in that way, he felt, and additional constraints were emphasized by other members of the committee. Tom York-King, a CLA nominee now in semi-retirement, saw Wookey's practices as embodying what he understood by the term "good husbandry", adding regretfully "It seems to become more obvious to me that that was the right way and not the methods of today". Intensive chemical farming creates a dependence on such programmes and Charles Blackwood, another CLA nominee, said that whilst he would "love to be able to do it that way" it was simply not an option for him because of the inevitable cost in making the transition.

Not surprisingly, perhaps, there was general enthusiasm for the Game Conservancy's Cereal and Gamebirds Project which was demonstrating how unsprayed headlands could make a contribution both to wildlife conservation and a less intensive approach to production. The project's steering committee Chairman, Hugh Oliver-Bellasis, had spoken at a meeting organized by Wiltshire FWAG, and the Adviser had visited the Game Conservancy's headquarters at Fordingbridge to learn more about the work. Conservationists on the FWAG committee similarly approved of the project but were, interestingly, somewhat more sanguine about Barry Wookey whom they tended to view as an "organic monoculturist", not only as committed to the sort of tidy farming which is ultimately inimical to wildlife and habitat diversity as any conventional farmer, but also inclined to include in his conception of conservation the planting – as Lesley Balfe put it – "of all sorts of exotic species which have little to do with Wiltshire".

Nevertheless, the general approbation and even, amongst the farmers, envious admiration for Wookey's activities indicated an incipient change of awareness forced, no doubt, as much by the uncertainty afflicting the agricultural industry as it had been fostered by active involvement with FWAG. Indeed, an increasing recognition that fundamental change was necessary was articulated in ways characteristic of the basic FWAG message: an appropriate balance, now missing, would have to be struck. Although the days were past of the adviser being phoned anonymously because the caller did not wish to reveal an identity until the adviser had been 'sussed out', the changed policy context also meant that the resources from the farming community, so essential to FWAG's self-image and advisory capability, would now be harder to elicit. Wiltshire FWAG, in common with the FWAG movement as a whole, finds itself confronting a situation characterized by contradictory possibilities and pressures. It is to such issues that we turn in our concluding chapter.

Table 7.5   Enterprise type and farm size by hectares of advisory visits made by Wiltshire FWAG

	Up to 20	*20 to 80*	*80 to 200*	*over 200*	*Total*
Arable	-	4	13	17	34
Sheep	5	9	2	-	16
Dairy	-	28	31	6	65
Mixed	4	15	39	59	117
Beef	9	18	7	1	35
Horticulture	6	-	-	-	6
Horses	4	1	1	1	7
Other enterprises	14	6	-	-	20
Hobby	31	6	2	-	39
Total	73	87	95	84	339

CHAPTER 8
An Uncertain Future

It is evident from the account we have presented that it would be a gross exaggeration to claim that FWAG has had a major impact on the course of conservation policy or politics over the past decade. Nevertheless, in view of the prominence it has been accorded in a succession of Ministerial statements, evidence to Select Committees, press releases from the NFU and the CLA and so forth, an untutored observer would be forgiven for supposing that FWAG has played nothing less than a pivotal role in reconciling the competing claims of the farming and landowning community and an increasingly vociferous and well-informed conservation lobby.

It is in that sense, and without in any way seeking to downplay its record of practical achievement in often difficult circumstances, that any overview must begin by recognizing FWAG's primary significance as a tangible expression of the principle of voluntary co-operation; the article of faith for the farming and landowning community which became so central a tenet of the 1981 *Wildlife and Countryside Act*. Indeed, so troubled were FWAG's earlier years that it often struggled merely to survive. A changed policy context, however, brought it to the centre of attention. Other countryside agencies, moreover, were obliged to develop appropriate stances towards an organization for which extravagant claims were being advanced.

FWAG's development, therefore, tells us much about the wider context within which countryside and land-use issues have developed, and particularly about inter-agency rivalry in a congested policy area. More specifically, its changing fortunes have themselves furnished a commentary on the politics and ideology of agricultural and environmental regulation. Its prospects, too, depend on the shifting balance of forces between the major countryside agencies within the overall context of the Government's developing policy towards farm conservation. Following an analysis of FWAG's articulation of the voluntary principle, therefore, we examine its evolving links with the two agencies, the Countryside Commission and ADAS, which have most influenced its present form and function and are likely to play a crucial role in determining its future.

The Voluntary Principle

The passage of the 1981 Act, growing awareness of the environmental damage associated with modern methods of agricultural production, concern over agricultural surpluses, these and other factors generated

An Uncertain Future

conditions peculiarly conducive to the emergence of a farming and conservation advisory agency enjoying the support of a farming lobby anxious to vindicate its claim that "goodwill and voluntary co-operation" would be sufficient to protect and enhance the conservation value of the farmed landscape. FWAG, itself the product of a longstanding presumption in favour of voluntary responses to problems of countryside management, was pressed into frontline service in the struggle to resist the encroachment of statutory controls over damaging agricultural activities. Its very existence enabled it to function as a sort of talisman, ritualistically invoked as the mechanism through which farmer and conservationist might engage in constructive dialogue. Indeed, Eric Carter's observation that, if FWAG had not already existed in 1981 it would have been necessary to have invented it, is wholly apposite. But it is an observation that rings more true in terms of the ideology and politics of the post-1981 situation than it does with regard to matters of ecology and land management.

During the past decade, moreover, the talisman has been buffed and polished by the sustained determination to bring the organization's advisory capability into line with the claims perennially advanced on its behalf. Certainly it has done much to promulgate the policy presumption articulated so single-mindedly by both the NFU and the CLA and enshrined as the defining characteristic of the 1981 Act. In that sense the reproduction of an ideology and the development of a particular style of operation have, in the case of FWAG, been mutually determining. The ideological aspects of symbol systems are of course apparent at different levels and forms of representation. They range from more or less systematic, coherent and well-articulated 'world views' through taken-for-granted elements which, though seldom articulated, are significant for the way they structure perceptions and guide practices, to the least coherent or self-conscious level where ideology presents itself as a form of unreflective and unquestioned 'common sense'.

So strong at each of these levels has the commitment to a responsive and voluntary strategy been on the part of its members, that FWAG has inevitably found itself cast in a legitimatory role as the farming and landowning community has sought to defend its autonomy in the face of the demand that controls over landscape change be introduced. Certainly, and notwithstanding the organization's ostensibly neutral stance, FWAG officers have in their public statements often found themselves springing to the defence of the farming community.

In a letter to the Countryside Commission in July 1987, Lord Melchett expressed his sense of exasperated astonishment at an earlier letter from

Eric Carter as FWAG's National Adviser, questioning the value and validity of the Commission's Monitoring Landscape Change Survey which had shown that hedgerows in England and Wales were disappearing at a rate of 4000 miles per annum. It is significant, Melchett retorted, that the rate was 37 per cent greater in the period 1980-85 than in the previous decade, for it had been since 1980 that FWAG had received most support, funds and praise from the NFU, the Ministry and the Commission itself. Articulating the suspicions of sceptical conservationists he judged that "Eric Carter has certainly confirmed that FWAG is prepared to be an apologist for the worst excesses of modern agriculture".

The voluntary principle, though, is two-sided: whereas, negatively, it is an ideological defence of the autonomy of farmers and landowners; positively, it is about encouraging a social ethic concerning stewardship of the countryside. Writing in 1977, in the year which saw the publication of *Caring for the Countryside*, Joan Davidson outlined three short-term priorities for the conservation of the farmed countryside: protection of the remaining natural or semi-natural habitats from agricultural intensification; the remedial management of agriculturally modified habitats to enhance their conservation value; and, finally, the creation of new elements to increase the ecological richness and scenic diversity of intensively farmed areas.

In practical terms, FWAG has tended to be associated principally with the last of these, even though it is arguably the least important from a conservation point of view. Such an approach has often been justified, of course, as an effective way of engaging the interest of the uncommitted farmer. In Norman Moore's characterization:

> FWAG is involved in a long-term plan of education, and experience has shown that the first and critical contact between itself and the farmer is usually provided by tree planting. FWAG advisers must respond to that request and build positively on that foundation . . . . One of the practical problems of conservation on the farm is that it takes time to get results. Therefore any conservation activity which quickly produces dividends is very welcome. (Moore, 1987, p.112)

It is easy enough to disparage such pragmatic realism as betraying little more than a concern with the cosmetics of conservation. To sceptics – and, as the Wiltshire Trust for Nature Conservation's Field Officer put it in interview, "everyone who is serious in the countryside is sceptical about

FWAG" – FWAG's overwhelming commitment to compromise has encouraged farmers to suppose that all they have to do for conservation is plant trees and dig out ponds. Its main concern, therefore, has tended to be with secondary habitats which are recreatable and the attendant danger is that farmers believe they are doing the right thing even as they continue to damage or destroy important primary habitats.

FWAG could hardly be other than acutely aware of this widely voiced criticism. It has responded by pointing out that it cannot be blamed if this is how individual farmers behave for it has made its own priorities very clear; that the maintenance of existing woods, for example, is more important than planting new ones (*FWAG Guide to Priorities: Wildlife on the Farm*, 1982).

It is evidently not the case, however, that these priorities have always been so clearly imparted to farmers. In diluted form, the FWAG ideology that nature conservation and commercial farming can happily co-exist may have confirmed some farmers in their belief that effective conservation need not involve making hard choices that demand some personal costs. In Shropshire, Norfolk and Suffolk FWAGs, Potter (1985 and 1986) found that typically the farmers who had been more than usually active in conservation improvement projects, such as tree planting and the creation of farm ponds, did so to make amends for previously damaging or over-zealous improvements. Of course, in itself this is no bad thing that someone is led to take remedial action out of a wish to atone for previous 'misdeeds'. FWAG's role must be to channel such impulses into constructive effort and a deeper perception and commitment. To do so effectively, it must be able to counter the charge that "FWAG is unable to lock into farmers' decision-making at an early enough stage and is consequently limited to amelioration and enhancement once 'improvement' has run its course" (Potter, 1985).

The force of the prevalent criticisms of FWAG relate to the standard of advice and support that it can give, and above all, perhaps, to how strategic is its input into farmers' decision-making. On retiring as the Chairman of national FWAG, Norman Moore made the following appeal for more attention to be given to the quality of advice on offer:

> Increasingly we shall be judged by what we achieve on the ground. I should like to make a strong plea for more professionalism. FWAG should become known for its association with really effective conservation management. If we condone poor management practices we give the impression that we are not serious, and we lay

ourselves open to the charge that we merely encourage cosmetic treatment of the countryside. (FWAG Newsletter January 1985)

The development of FWAG's advisory capacity during the past few years, through the appointment of county advisers, has gone a good way to meet the criticisms. The initial and quite understandable concern with the quantity of advice given should now give way to the quality of support to achieve tangible conservation gains. This is likely to be best done on a whole-farm basis and before farms are subject to extensive rationalization. Return visits over an extended period may be required to advise on implementation and to monitor the results.

Undoubtedly, FWAG advisers are highly motivated and highly qualified, and, with the above caveats, the commitment of the Countryside Commission and the Farming and Wildlife Trust to such an advisory programme must be judged a major advance in farm conservation policy. FWAG itself, meanwhile, has generated a clearer sense of its own identity as well as an account of its genesis not entirely free of the impulse to mythologize.

Speaking after dinner on the first day of the National FWAG Conference held in 1986 at the Royal Agricultural College, Sir Derek Barber presented a broad overview of that development with the title 'Silsoe to Cirencester and Beyond'. Beginning with an account of the development of FWAG which avoided any reference to the substantial difficulties which had been encountered, he paid tribute to Jim Hall's "enormous enthusiasm" and then characterized, in the most self-effacing manner imaginable, the period during which FWAG's credibility had been substantially bolstered. Then, in the briefest of possible characterizations of a complex episode he simply said, "the Countryside Commission got interested and produced quite a lot of money". In fact, as we have seen, the form and character of the Commission's involvement looked likely to do nothing less than transform the impact FWAG might have on the management of the farmed countryside.

FWAG and the Countryside Commission: the Tension between Client and Sponsor

One of the Commission's key objectives in following up its experimental work with demonstration farms and countryside management projects had been to ensure that those wishing to carry out conservation measures could receive appropriate advice and assistance. The Countryside

Adviser Scheme, launched in 1979, envisaged such appointments effecting an integration of resources for countryside conservation such that an adviser would, by drawing upon specialist sources, become the focal point for advice over a wide area. Discussions between FWAG and the Countryside Commission prior to setting up the Farming and Wildlife Trust, however, indicated a number of points which have remained contentious. For, although the Commission considered Countryside Adviser the appropriate designation for appointments under the scheme, and continues to refer to FWAG advisers in this way, FWAG itself gives them the title Farm Conservation Adviser.

The difference is far from trivial and symbolizes a number of factors about which the Commission still has misgivings. They envisaged advisers fulfilling a 'gatekeeper' role, acting as first point of contact and calling in specialist knowledge. Things have, however, typically developed in a different way and conservation bodies have felt somewhat separated from FWAG advisers. Partly this stems from the policy of the National Adviser who, keen to build a strong FWAG image, presented FWAG as able to respond to any request for help. But it also reflects the fact that at the time when many FWAG posts were coming on stream the efforts of the NCC, for instance, were necessarily almost wholly directed towards their SSSI designation responsibilities in the wake of the 1981 Act.

The Commission's Countryside Advisers were, moreover, expected to be as concerned with the landscape, recreation and access isues as with wildlife conservation. Indeed, advisers were to be required to attend a course on landscape conservation approved by the Commission within 12 months of their appointment. This condition was insisted upon in relation to FWAG advisers and there was, at first, considerable resentment over what was seen as an unnecessary imposition. Courses were shared initially with Countryside Rangers and others, who typically evinced a very different attitude towards questions of access, and this only exacerbated the discontent. The Countryside Commission's Demonstration Farm Projects were also very much concerned to show how recreation and conservation could go hand in hand. But the Commission has consistently failed in its attempts to elicit FWAG's help in relation to footpath problems, for instance, and there is a feeling within the Commission that FWAG is all too ready to mouth the NFU attitude to such issues. These tensions are far from resolved, and FWAG retains its fundamental wildlife orientation.

If such matters give rise to misgivings within the Commission there is disquiet too concerning FWAG's continued 'arms length' relations with local authorities: notwithstanding the fact that the advisers in Avon and Bedford are actually employed by them and a number of others are based in council offices. Local authorities still have no representation on national FWAG and the often apparent anti-planning prejudice against them within the FWAG committee structure is a matter of embarrassment to the Commission which has always worked very closely with local councils. Most of the Commission's initiatives following the New Agricultural Landscapes study were predicated upon the close co-operation of local authorities, often acting as agents for grant-aid schemes. As the Commission's 1979 plan made clear, its ambition was to establish a whole range of countryside management projects as an adjunct to countryside planning. But such hopes were effectively stymied by the changed situation which local authorities confronted after 1979.

At a time when local authorities were starved of funds and facing an uncertain future, the number who could take an interest in the Commission's schemes was limited. FWAG appeared to be in a position to respond precisely at a time when local authorities could not. Indeed, the prominence accorded to FWAG and the support it has enjoyed, if not entirely fortuitous, has been substantially enhanced by the passage of the *Wildlife and Countryside Act* 1981, its workload implications for the NCC and the difficulties faced by local authorities in the wake of the Conservative victory in the 1979 general election. Given its hopes for the Demonstration Farm Projects and other initiatives it is easy to see why enthusiasm for FWAG within the Commission was so qualified and why the decision to back FWAG took some considerable time to reach.

Crucial to its support, of course, has been the impetus provided by Sir Derek Barber's commitment. He often, as it were, appeared to hope that the reality would soon catch up to match the advanced billing he was giving this publicly funded, non-state organization: itself the very embodiment of the sponsored voluntarism which has characterized conservation policy over the last decade. Thus, presenting the judges' report on the 1984 Farming and Wildlife Award in *Country Life* (14 February 1985), he commented on the rapidly growing need for sophisticated advice and followed it, at a time when FWAG had appointed just 10 advisers, with the observation that "it is clear that the FWAGs have now come of age and represent possibly the most potent force in the whole conservation field".

The Commission's Chairman has been more reticent of late and the Commission itself having devoted considerable financial and organizational resources to FWAG, has been critically appraising its involvement. Other agencies, such as the NCC, have not put such resources into countryside advice. The hope that a network of countryside advisers could be established has been the Commission's, and circumstances have determined that the effort has been principally channelled through FWAG which has, thereby, enjoyed remarkable support from the agency. Indeed, it might be said that at a crucial moment in their respective development the two organizations, the one a statutory agency and the other an instance almost of 'private interest government' within the conservation field, needed each other.

The Commission's expectation of one adviser per county is the FWAG ambition as well, but the Commission made it clear that FWAG could not 'expect' any funding commitment beyond six years. The Commission which, as we have seen, cannot be wholly at ease with the way things have developed, recognizes that the FWAG experience is a very varied one and it is determined to treat each case on its merits. The Commission will continue to assert its view that the countryside adviser role should be a non-specialist one embodying wildlife and landscape concerns to an equal degree. There will also be greater emphasis in future on applications for funds being presented in terms of specific, measurable objectives. Moreover, if work on the ground is seen as the appropriate focus for the Commission's funding efforts, there is no reason why funding should be tied to particular posts.

The review of the FWAG experience is likely to mean that some counties will be given 'top-up' payments whilst others which are floundering badly in the fund-raising stakes will be carefully scrutinized before being given any extra help. Ultimately, the Commission is prepared to withdraw its backing even if this means the collapse of part of FWAG's present advisory network. Given the extent of its investment in the organization, this would be a painful decision but it is a nettle the Commission would be prepared to grasp in view of the many other worthy causes deserving its support.

Any overall appraisal of FWAG is bound to consider the adequacy of its past performance and present operating structure. But it will, additionally, be sure to have regard to the substantially modified policy context in which FWAG now operates. In a situation which proffers diverse sources of advice for farmers and landowners, FWAG clearly needs to define a distinctive niche for itself. Finding such a niche will be likely to entail a more positive alignment with the Countryside Commission's conception of its role as well as the development of a structure to provide scope for the career development of advisers and a more accountable framework for their activities.

FWAG and ADAS: Partnership or Rivalry?

There have been major upheavals in the agricultural policy and advisory world in the past few years and more are in prospect. None presents a greater challenge to FWAG than the changing role of ADAS. We have already noted that key individuals from the service were crucial in the emergence of FWAG and in its survival at the national level during its difficult formative years. But the 1980s have seen the development of a closer institutional relationship between the two with some inevitable tensions.

The 1981 *Wildlife and Countryside Act* required ADAS to give conservation advice to farmers requesting it. The most tangible and immediate consequence was increased support and prominence for FWAG in the advisory and promotional work of ADAS. This took a number of forms: office and secretarial support; allocation of staff time to FWAG business; promoting FWAG in ADAS-sponsored exercises and exhibitions; and the development of formal liaison and referral arrangements. It seemed that ADAS and FWAG, in amicable partnership, were putting themselves at the forefront of attempts to promote education and advice on farming and conservation. In the main, the direct involvement of ADAS officers in conservation problems would be concentrated on sensitive areas, such as SSSIs and National Parks. FWAG was to play the leading role in the wider countryside with ADAS devoting a small but significant proportion of its overall resources to backing FWAG.

A memorandum of agreement of September 1985 between the Director General of ADAS and the FWAG National Adviser (himself a former Deputy Director General of ADAS) seemed to set the seal on the partnership, laying down guidelines for the co-operation of the two organizations. ADAS recognized "the role of FWAG as a forum for discussion and as a source of practical advice and assistance on countryside conservation" and committed itself to continue to support the group's aims and objectives and to provide county FWAGs with back-up facilities. For its part, FWAG accepted the need to give due recognition and publicity to ADAS's contribution to its activities, and to liaise with Divisional Surveyors, where possible in advance, when FWAG advisory activities might impinge on designated land or sensitive sites. The two organizations agreed that local FWAG steering committees would discuss with their ADAS representative the range of specialist skills available in the locality and how advisory efforts could be mounted "in the best interests of agriculture and countryside conservation".

Recent political developments, however, have threatened the equilibrium of this partnership. Section 17 of the *Agriculture Act* of 1986 placed upon the Minister for the first time the duty to endeavour to achieve "a reasonable balance" between agriculture, the rural economy, conservation and the enjoyment of the countryside by the public when implementing policy relating to land. It also provided for the designation of Environmentally Sensitive Areas, in which the Ministry would seek agreements with farmers to farm in ways beneficial to conservation: for the first time, conservation objectives were to be pursued as a direct and integral feature of agricultural support. Another part of the Act with implications for advisory policy was Section 1 which enabled the introduction of charges for ADAS advice. Ministerial assurances were given throughout the passage of the bill that conservation advice would remain free although this has been interpreted as applying only to general, and not specific advice. These legislative changes have brought conservation on to the centre stage of ADAS activity, and it is evident that a prominent element of ADAS's work in the future will lie in the provision of conservation advice and in the implementation of conservation policies. Already, such policy developments as the Farm Diversification Scheme, the Farm Woodland Scheme, Set-Aside and the recently revised Farm Grant arrangements have given the service a much enhanced role with regard to conservation in the farmed landscape. This has thrown ADAS's relations with FWAG into uncertainty. Eric Carter, the then National Adviser for FWAG, commented, "Really it is a very difficult area. We would like to have more positive co-operation, but at the moment it is very vague; they [ADAS] don't seem to know themselves" (*Countryside Campaigner*, Autumn 1986). The developing partnership had been premised on conservation being peripheral to ADAS's main concerns. Now there is the possibility of serious competition between the two organizations, though ADAS has so far largely confined its role to specific schemes.

The capacity of ADAS, as presently constituted, to provide a major farm conservation advisory service has seriously to be questioned, however. First, there is the problem of image. As one farmer put it: "How can ADAS officers start getting people to conserve, say, ten per cent of their old pastures when it's spent the last twenty years encouraging them to plough them out?" (quoted in Pye-Smith and North, 1984). The history of the service shows a preoccupation with narrowly defined production issues and a retreat from the aim of supplying advice to any individual farmer seeking it to that of a general information service of greatest use to the larger and more sophisticated farms. As a result many farmers have had little or no direct contact with ADAS. By the same token ADAS officers have not developed the complex extension skills required in countries where the 'farm problem' has required an advisory service with a more active rather than a responsive role. It is clear that the promotion of conservation policy will require a pro-active, innovative and proselytizing approach if it is to reach the majority rather than a minority of farmers, and if it is to affect everywhere the generality of farming practices. The slow and piecemeal development of socio-economic advice since the mid-1970s (Woods, 1980) does not augur well for the future of conservation advice. Similarly the 'commercialization' of other aspects cannot improve resources available to the advisory service. Its ability to respond is currently limited enormously by staff cuts, lack of suitably trained advisers and minimal opportunities for substantial re-training or the recruitment of specially trained advisers.

Nevertheless, to the extent that conservation policy for agriculture moves beyond providing information and promoting voluntary co-operation towards a more regulatory approach (as in SSSIs) or one dependent upon grant inducements (as in ESAs and Set-Aside) that requires direct government agency, it is unavoidable that ADAS assumes a central role. Indeed, it is not so improbable that ADAS's major function in the future could be farm conservation and socio-economic advice and regulation, with advice relating solely to productivity matters being largely catered for by private consultants and the representatives of the supply industry. Without major reform of its organization and personnel, however, there must be considerable doubt about the quality of the conservation advice that ADAS can offer.

If it is to become an effective agency for promoting integrated farming and conservation it will need guidance, support and stimulus and here its relations with FWAG could be of crucial importance. In many respects, FWAG stands well within an honourable tradition of British public life whereby a voluntary organization – albeit, one with significant links to statutory organizations – identifies and responds to an important, unmet social need. In many cases the voluntary initiative is superseded once the state assumes responsibility for the matter. In some instances, for example when the Sports Council was established, the state has actually taken over the structure of provision previously established by the voluntary organization. In this case, the situation is somewhat different, calling not for the establishment of a new state agency but the reform of an existing one. This could well be the next major task in which FWAG is involved. A close partnership between ADAS and FWAG would help the former through the difficult transition period ahead.

FWAG can bring some crucial strengths to this partnership. As a comparatively fresh organization in the advisory field it has a pioneering vigour not present in ADAS. It has not had to re-train agricultural staff but has been able to recruit specialist advisers who are usually extremely well qualified in ecology and conservation. They are not hide-bound by custom or by fiat. FWAG is independent and rooted in an organization which brings together farmers and conservationists and which, in the words of Wilf Dawson, the Director of the Farming and Wildlife Trust, is "farmer motivated and to a considerable extent, farmer-managed" (*Countryside Campaigner*, Autumn 1986). This is an important factor given the widespread disillusion amongst farmers towards the official advisory service.

On its own, though, FWAG could never provide a full-blown farm conservation advisory service, even if it was desirable that it should. It has neither the resources nor the statutory authority nor the policy implementation role nor, indeed, does it have the range of contacts with farmers and landowners enjoyed by ADAS. After all, ADAS officers until recently outnumbered FWAG advisers by 100 to 1. For this reason, if no other, amalgamation of FWAG with ADAS would not be a satisfactory solution, as it would be very difficult for FWAG to retain its identity were it to be subsumed into the larger organization. Instead, FWAG's role must be that of the external catalyst.

How might this be strengthed? There seem to be at least four desiderata. The first is that ADAS staff should be exposed to the lessons and insights accumulated by FWAG and its advisers: there must be an efficient transference of experience, know-how and skills. The second is that ADAS needs to renew both its links with and its image within the farming community, and here FWAG could play a vital mediating part. The third is that ADAS must become more responsive to local and national conservation authorities. Fourth, and finally, if ADAS assumed the central position in giving farm conservation advice, FWAG's promotional, innovatory and mediatory functions should be reinforced. A crucial issue is clearly that of providing an appropriate structure to achieve the necessary integration. It is questionable whether FWAG county committees as currently constituted are adequate for the task. Whilst they have been extremely valuable in forging necessary links between the conservation agencies and leaders in the farming community, they have neither the constitutional powers nor the administrative resources to play a greater role in this sphere.

Some reorganization of the Farming and Wildlife Trust's operations is called for so that it might more effectively act as first point of contact and co-ordinator of a spectrum of advice in the manner envisaged by the Countryside Commission. Were FWAG capable of evolving a structure whereby boards established at the regional level might be given responsibility for administering grant aid, then a number of other desirable developments might be simultaneously encouraged. Each regional FWAG board would find itself employing a number of advisers from a central office and this would, in turn, create a basis for developing a career structure since a number of adviser grades could be created. Young and inexperienced advisers, particularly, would benefit from the existence of a structure to which they could relate and there would be administrative and procedural benefits as well. Presently advisers often find their loyalties divided between independent committees, the national Farming and Wildlife Trust and even, to some extent, the institutions in which their offices are based.

FWAG Prospects

The prospects for some such development are, of course, dependent on prior resource commitments. Indeed, the most pressing issue, if FWAG is simply to sustain, let alone expand, its present level of activity is that enhanced and secure funding be found. Otherwise, the gradual withdrawal of Countryside Commission funding may precipitate the sort of hopelessly weakening haemorrhage which has often threatened to interrupt FWAG's progress in the past.

Fund raising in a competitive environment is never easy but FWAG's track record in this respect is bound to give cause for disquiet. Past experience and present evidence each suggests a very patchy response from the farming community and its allied industries. Already, before the current wave of funding from the Farming and Wildlife Trust and the Countryside Commission, some of the groups that had pioneered the appointment of advisers found themselves confronting this problem. Thus, as the initial grant aid for their Adviser diminished, Gloucestershire FWAG reported that "An appeal for funds has been sent to the Adviser's clients and they have responded generously, but have stressed that they do not feel this is the correct way to sustain the post" (1982). When the national scheme for supporting FWAG advisers was introduced, Gloucestershire FWAG experienced difficulty in raising the matching 25 per cent needed to maintain the service. A subscription scheme attracted fewer than 5 per cent of the county's farmers and a voluntary levy was not a success. Even some of the farmers who had taken advantage of the service and had obtained grants as a result, failed to contribute (*Gloucestershire Agricultural Journal*, 16 May 1985).

In the meantime, the downturn in the economic fortunes of the agricultural industry has exacerbated the problem. Thus, whereas a few counties, whose members are well connected and capable of bringing a certain flair to the exercise, are relatively secure, others have been unable to move towards the sort of self-sustaining financial independence envisaged for them. That, in turn, has meant that in many instances advisers are acutely aware of their own insecurity. And whilst an element of uncertainty may be conducive to effective performance its more thoroughgoing variant is likely to lead to an inappropriate diversion of energies into fund raising and, at worst, to a wholly unproductive demoralization.

Morale could only have suffered further when, in 1989, in an indication of the seriousness of FWAG's financial state, a halt was called to further recruitment of advisory officers, even to replace any who resigned, leading to a fall in their number (Figure 4.1). Dorset, for example, had to revert to offering an advisory service based on the volunteer members of the county committee. In an equally regrettable move, charges were introduced for advisory visits. Not only did this overturn a strong commitment amongst many advisers and their committees to the principle of a free service but it also put them potentially in a competitive relationship with a range of agencies in both the private and public sectors. The commercialization pressures on ADAS, for example, had already led to some questioning of the free administrative support given to some FWAG advisers by ADAS offices, and in isolated cases to its curtailment. The prospect of charges, though, threatens to place ADAS and FWAG in a directly competitive position and to strain relations even further. Moreover, it is uncertain how farmers will respond to the introduction of charges.

FWAG's financial situation, therefore, would seem to be precarious. Nevertheless, it must be said that there are many features of the present situation which are conducive to the recognition of an obligation to sustain FWAG. We are witnessing the first concerted efforts in recent decades to adjust the direction of agricultural development in Western Europe. Following four decades of 'productivist' agricultural policy, driven primarily by the need to raise productivity and improve farming efficiency, over-production, falling farm incomes and the rising budgetary costs of farm support have called into question the assumptions which lie behind the post-war model of agricultural technology and policy. At the same time, and partly in response to the adverse consequences of this approach, pressure for improved environmental regulation has grown and a number of recent policy developments have signalled a recognition of the need to reduce farming's use of resources and make its operation more accountable.

Farmers and their families will increasingly be seen as managers of countryside change. But the policy presumptions which have characterized the 1980s have, from the 1981 *Wildlife and Countryside Act* onwards, overwhelmingly favoured a voluntary approach which makes only minimal recourse to formal controls. Although the Government has been obliged to adopt a more interventionist stance it has retained a commitment to a non-directive policy style which accords a high priority both to property rights and the discretion which can be exercised in the matter of policy implementation. In this context FWAG plays a significant role: not least because it can be referred to as an appropriate source of advice for farmers choosing to implement one or more of the schemes increasingly on offer.

Participation in the Set-Aside and Environmentally Sensitive Areas schemes, for example, is voluntary in each case, and this despite the Government's insistence that they constitute an essential basis for the re-orientation of agricultural and environmental policies in the UK.

In the case of Set-Aside the principle of voluntarism is extended still further in that although the Government has made clear its wish that the scheme be used for positive environmental purposes it has, nevertheless, substantially left the achievement of such purposes to the managers of the land. The voluntary principle is similarly to the fore in relation to non-agricultural measures such as the Farm Woodland Scheme. Though it cannot be said that FWAG played any significant role in the emergence of these policy initiatives, they must necessarily be accompanied by an effective promulgation of the voluntaristic ethic, and it is in relation to the reproduction of that ideology that FWAG could play an increasingly important role.

As we have indicated, it fulfils at the most minimal level a legitimatory role merely by being available for presentation as a cornerstone of the present policy strategy. In that sense – and as we have made clear it is by no means the only reason – there will be a great reluctance on the part of the agricultural policy community to see it falter seriously. So closely has it been associated with the voluntaristic philosophy that for the Trust to fail would be to call into question the approach itself, since the FWAG groups and advisers are the most readily accessible way in which farmers can demonstrate their commitment to goodwill and voluntary co-operation.

FWAG's identity is arguably of greater importance now than ever before as farmers adjust their perceptions of their commitments and actively seek to produce conservation. Farmers, overwhelmingly, see conservation activity in terms of taking positive actions and, whilst there is scope for pointing to the inadequacy of a view which often fails to appreciate the more important value of refraining from damaging activities, it is certainly the case that FWAG has done much to encourage the establishment of specific features and the sensitive management of existing ones. It would surely be folly to squander the commitment that has been so painstakingly nurtured during often combative times.

Indeed, such is the resonance of FWAG ideology with the dominant presuppositions of present policies that the announcement in April 1989 of the award of a grant of £180 000 over three years jointly by the DoE and MAFF to the Farming and Wildlife Trust could well be the prelude to a more overt acceptance of Ministerial responsibility for the future well-being of the Trust. In a speech to the Oxford Farming Conference, delivered on the same day as the exhibition marking the beginning of Food and Farming Year had been opened, the Minister for Agriculture presented an overview notable for its recognition of the necessity for fundamental change in agriculture. But the talk, titled 'Agriculture – Just Another Industry?' was not wholly preoccupied with questions of structural transformation. The opportunity was also taken to advance the claim that recent legislative changes and policy adjustments make it appropriate, once again, to reassert with only minor modification the conventional wisdom of the post-war period which effortlessly presented farmers as the custodians of the countryside.

Increasingly, he suggested, conservation is being combined with efficient and competitive farming: "the work of the Farming and Wildlife Advisory Groups, the tremendous response to the Environmentally Sensitive Areas Scheme, the interest being shown in the Farm Woodlands Scheme and the take-up of grants for the reconstruction and maintenance of farm walls and hedgerows all demonstrate, if this were needed, the commitment of the industry to its custodianship role". FWAG is one critical element in a process of ideological renewal which reaffirms the central tenet to which it has been consistently committed. The government grant has assisted in the appointment of additional farm conservation advisers, thus reversing the decline in their number. By August 1990 there were 43 in post (31 in England, 1 in Northern Ireland, 8 in Scotland and 3 in Wales).

Forced to venture an opinion, even the most circumspect analyst would be bound to acknowledge that the fundamental logic of the situation in which the Trust now finds itself speaks loudly in its favour. Too much has been invested in FWAG for it to be allowed to stumble just at the moment when policy developments seem to have opened a space within which it could become more effective than at any previous time in its often fitful history.

APPENDIX

FWAG Information Leaflets

What Can I Do as a Farmer?*
Wildlife on the Farm – A FWAG Guide to Priorities
Do You Have a Resident Barn Owl?
When Did You Last See an Otter?
Trees Without Tears
Take a Pride In Your Landscape
Trees, Plants and Ponds
The Conservation of Frogs and Toads
A Nature Reserve and a Game Covert
A Guide to Planting Trees & Shrubs to form a Wildlife Area*
Scrub Management
Farm Woodland
A Hedgerow Code of Practice*
Conservation on Tenanted Farms
Pond Management*
Butterflies on the Farm
What Can I do as a Contractor?
The Conservation and Management of Old Grassland
Roadside Verges: A Guide to Users and Managers
Trees on Farms in the Lowlands
Creating Wild Flower Meadows*
Countryside Management – Sources of Help & Advice

* in print, August 1990

References

Advisory Council for Agriculture and Horticulture in England and Wales (1978) *Agriculture and the Countryside* (the Strutt Report), MAFF, London.

Barber, D., ed. (1970) *Farming and Wildlife: A Study in Compromise*, RSPB, Sandy.

Barber, D. (1971) 'The farmer and conservation – a compromise', *New Scientist*, 29 April, 270-271.

Barber, D. (1983) *The Chalkland Exercise on Chalkland Farming and Wildlife Conservation*, MAFF, London.

Barber, D. (1986) 'Silsoe to Cirencester and beyond'. Chapter 5 of *Managing Change: Report of the National FWAG Conference*, Farming and Wildlife Trust, Sandy.

Beynon, N.J. (1984) *Conservation Teaching in Agricultural College Courses: A Case Study of the National Certificate in Agriculture*, Trent Papers in Planning No. 21, Trent Polytechnic, Nottingham.

Blunden J. and Curry, N., eds., (1985) *The Changing Countryside*, Croom Helm, London.

Brookes, S.K., Jordan, A.G., Kimber, R.H. and Richardson, J.J. (1976) 'The growth of the environment as a political issue', *British Journal of Political Science*, 6, 245-255.

Carter, E. (1979) 'Ploughing a new furrow', *Birds*, 7, No.7, 21 - 25.

Centre for Rural Studies (1990) *Review of the Countryside Commission's Countryside Advisers Policy*, Royal Agricultural College, Cirencester.

Cornwallis, R.K. (1968) 'Change on the farm', *Society for the Promotion of Nature Reserves Handbook*, pp. 22 - 27.

Cornwallis, R.K. (1969) 'Farming and wildlife conservation in England and Wales', *Biological Conservation*, 1, 142 - 147.

Corrie, H. (1984) *Conservation Advice to Surrey Farmers*, MSc thesis in Conservation, University College London.

Country Landowners' Association (1982) *Planning and the Countryside*, CLA, London.

Countryside Commission (1974) *New Agricultural Landscapes*, The Commission, Cheltenham.

Countryside Commission (1982) *Countryside Issues and Action*, The Commission, Cheltenham.

Countryside Commission (1984) *Agricultural Landscapes: A Second Look*, The Commission, Cheltenham.

Countryside in 1970 (1964) *Proceedings of the Study Conference*, HMSO, London.

Countryside in 1970 (1966) *Second Conference*, Royal Society of Arts, London.

Countryside in 1970 (1970) *Third Conference*, Royal Society of Arts, London.

Countryside Review Committee (1978) *Food Production in the Countryside*, HMSO, London.

Cox, G. (1981) 'Wookey's Way Without Chemicals', *The Field*, 2 December 1981, 1198 - 1201.

Cripps, J. (1979) *The Countryside Commission. Government Agency or Pressure Group*, Town Planning Discussion Paper 31, University College London.

DART (1981) *Country Farming and Wildlife Advisers.* An unpublished report to the Countryside Commission.

DART (1983) *Small Woods on Farms,* Countryside Commission, Cheltenham.

Davidson, J. and Lloyd, R., eds. (1977) *Conservation and Agriculture,* Wiley, Chichester.

Dorset Naturalists' Trust (1970) *Farming and Wildlife in Dorset, Report of a Study Conference,* DNT, Poole.

Fairbrother, N. (1970) *New Lives, New Landscapes,* Architectural Press, London.

Gloucestershire FWAG (1982) *Gloucestershire Farming and Wildlife Adviser 1979 to 1982.* Gloucestershire FWAG, Gloucester.

Green, B. (1971) 'Countryside conservation and agriculture', *Rural Life,* 6 (4), 27 - 30.

Green, B. (1975) 'The future of the British countryside', *Landscape Planning,* 2, 179 - 195.

Green, B. (1981) *Countryside Conservation,* Allen and Unwin, London.

Heaton, A. (1984) *The Trusts and FWAG.* Unpublished results of a survey conducted by the Royal Society for Nature Conservation.

Keenleyside, C. (1977) 'Voluntary action in conservation', pp. 114 - 71, of Davidson and Lloyd, *op. cit.*

Kerby, G. and Hastings, M. (1987) *A Training Needs Assessment for FWAG Advisers,* Agricultural Training Board.

Leefe, J.D. (1968) 'Tree planting on the farm', *Lindsey NAAS and APC Farming Bulletin,* no. 250, May.

Leefe, J.D. (1970) *Lindsey Project for the Improvement of the Environment,* Lindsey and Holland Rural Community Council, Lincoln.

Lowe, P., Cox, G., MacEwen, M., O'Riordan, T., and Winter, M. (1986) *Countryside Conflicts: The Politics of Farming, Forestry and Conservation,* Gower, Aldershot.

Lowe, P. and Flynn, A. (1989) 'Environmental policy', in J. Mohan (ed) *The Political Geography of Contemporary Britain,* Macmillan, London.

Lowe, P. and Goyder, J. (1983) *Environmental Groups in Politics,* Allen and Unwin, London.

MacEwen, A. and MacEwen, M. (1987) *Greenprints for the Countryside? The story of Britain's National Parks,* Allen and Unwin, London.

Mathias, J. and Jolliffe, W. (eds) (1974) *The Cowbyers Conference on Upland Farming, Forestry, Game Conservation and Wildlife Conservation,* MAFF, Newcastle upon Tyne.

Ministry of Agriculture, Fisheries and Food (1973) *The Chalkland Exercise,* ADAS.

Ministry of Agriculture, Fisheries and Food (1975) *The Dinas Conference Report,* ADAS.

Ministry of Agriculture, Fisheries and Food (1976) *Wildlife Conservation in Semi-Natural Habitats on Farms,* HMSO, London.

Moore, N.W. (1987) *The Bird of Time,* Cambridge University Press.

Munton, R., Marsden, T. and Eldon, J. (1988) *Occupancy Change and the Farmed Landscape,* CCD 33, Countryside Commission, Cheltenham.

Nature Conservancy Council (1977) *Nature Conservation and Agriculture,* NCC, London.

Potter, C. (1985) 'From little acorns? The role of FWAG in the conservation of lowland landscapes', *ECOS*, **6** (3), 6 - 10.

Potter, C. (1986 a) 'Processes of countryside change in lowland England', *Journal of Rural Studies*, **2** (3), 187 - 195.

Potter, C. (1986) 'Investment styles and countryside change in lowland England', pp. 146 - 159, in Cox, G., Lowe, P. and Winter, M., eds. *Agriculture: People and Policies*, Allen and Unwin, London.

Ratcliffe, A.J.B., ed. (1972) *The Dinas Conference on Upland Farming, Forestry and Wildlife Conservation*, ADAS/MAFF, London.

Reynolds, S. (1984) *County Farming and Wildlife Advisory Groups*, Dartington Institute, Totnes.

Rowlands, E. (1972) 'The politics of regional administration: the establishment of the Welsh Office', *Public Administration*, Autumn, 333 - 351.

Sandbach, F. (1980) *Environment, Ideology and Policy*, Basil Blackwell, Oxford.

Shoard, M. (1980) *Theft of the Countryside*, Temple Smith, London.

Silsoe Report (1969) *Agriculture and Nature Conservation Study Conference*, RSPB, Sandy.

Sly, A. (1978) 'Wildlife and the farmer', *NFU Insight*, **116**, 8.

Stubbs, J. (1977) *The Concept of SSSIs: A Review*. Unpublished thesis, University of Southampton.

Suggett, R.H.G. (1982) 'Conservation in Agricultural Education Guidance Group', DES Teachers' Short Course NZ, 13 - 16 September, 1982, Warwickshire College of Agriculture.

Tinker, J. (1969) 'The battle of Pendley Grove', *New Scientist*, 24 July, 194.

Winter, M. (1985) 'Administering land-use policies for agriculture: a possible role for county Agriculture and Conservation Committees', *Agricultural Administration*, **18** (4), 235 - 249.

Winter, M. (1988) 'Information and advice to Farmers: the development of conservation advice for farmers in England and Wales', in *Environment and Agriculture*, Proceedings of European Year of the Environment Seminar, Aslon, Belgium. pp. 163 - 170.

Wookey, B. (1987) *Rushall: The Story of an Organic Farm*, Basil Blackwell, Oxford.

Index

Access to the countryside, 179
Advice, nature of requests for, 97-102
Advisers
   see countryside advisers;
       farm conservation advisers;
       Farming and Wildlife
         Adviser;
       local groups (FWAG);
       qualifications and experience
         of advisers
Advisory Council for Agriculture and
   Horticulture, 40-1
Advisory leaflets, 11-12, 98, 101, 190
   see also Farming with Wildlife
Advisory service, national
   see national advisory service
'Advisory Services', 68
Afforestation, 83
Agricultural Development and
   Advisory Service, (ADAS) 4, 5, 25,
   28, 30, 40-1, 53, 66, 69, 77, 86-88, 91-3,
   103, 110, 112, 113, 116, 117, 120-2, 132-3,
   134, 138, 151, 155-6, 164, 174, 182-6
   Agriculture Service, 122
   Land and Water Service (LAWS),
      119, 122, 145
   Wildlife and Storage Biologists, 96
Agricultural development in Western
   Europe, 187-8
Agricultural Executive Committee, 11
Agricultural Habitat Liaison Group,
   36, 38
Agricultural Land Service, (MAFF) 12
Agricultural shows, r77, 103
Agricultural Training Board, 164, 167
Agriculture, changes in, 13, 36-8, 50
Agriculture, compromise plan for,
   16-17, 28-9
Agriculture, intensification of, 16, 83
Agriculture Act, 64, 183

'Agriculture and the Environment',
   64, 65
Amenity societies, 8
Andrews, John, 44-5
Arable farming, 83
Archaeological Officers, County, 95
Archaeological societies, 90
Areas of Outstanding Natural Beauty,
   111, 118, 124, 129, 142
Avon County Council, 94, 180

Badger Working Group *see* Wildlife
   Link
Badgers, 157-9
Balfe, Lesley, 145, 159, 170
Barber, Sir Derek, 13, 14, 15, 16, 19, 20,
   27, 28, 29, 31, 46, 49, 56, 59, 67, 113,
   147, 178, 180
Batten, Leo, 30
Bedfordshire County Council, 94
Bedfordshire FWAG, 100, 105, 180
Berwyn SSSI, 112, 120-1, 123
Beynon, John, 162
Biological Conservation, 11
Bishop Burton College of Agriculture,
   72-3, 161
Blackwood, Charles, 145, 169
Bovine tuberculosis, 157
Brecknock FWAG, 122
British Association for Shooting and
   Conservation, (BASC) 89-90, 95
British Farmer and Stockbreeder, 38, 40,
   51
British Field Sports Society, 90
British Hedgehog Preservation
   Society, 96
British Trust for Ornithology, (BTO) 2,
   13, 14, 28, 29, 33, 86
Brown, Col. Jack Houghton, 130, 135
Butler, Sir Richard, 59, 60

Cadbury Trust *see* W.A. Cadbury Trust
Cambridge University Department of Land Economy, 20
Caring for the Countryside, 39, 40, 132, 133, 176
Carmarthenshire FFWAG, 117, 122
Carnegie United Kingdom Trust, 10
Carter, Eric, 12, 15, 20, 28, 31, 35, 38, 47, 53, 81, 91, 113, 117, 146, 153, 175, 176, 183
Castle Caerinion, 113, 124
Centre for Rural Studies, Royal Agricultural College, 86, 104
Ceredigion FFWAG, 103, 117
Chadacre Farm Training Centre, 70
Chalk grassland, 129
Chalklands Exercise, 27, 28, 29, 30-1, 129-132, 134-6, 144
Cheshire, 27, 30
Cheshire FWAG, 90
Clarke, Gordon, 77
Cobb, Paul, 59
Cobham, Ralph, 134-5
Code of Good Agricultural Practice, 96
Coed Cymru, 119
Colleges of Agriculture, 77, 86
  see also Centre for Rural Studies
    individual Colleges
Common Agricultural Police, 114
Community Council for Wiltshire, 131, 154
Compromise plan
  see agriculture, compromise plan for
Conder, Peter, 13, 20, 33
Conservation
  see county trusts for nature conservation;
    farmers;
    Farmers' Union of Wales;
    voluntary principle in conservation

Conservation advice
  see Agricultural Development and Advisory Service;
    farmers;
    Farming and Wildlife Advisory Group;
    Wiltshire
Conservation adviser, Wiltshire
  see Wiltshire (advisory team and Adviser)
Conservation advisers, 49, 55, 60
  see also Countryside Adviser Scheme;
    Wiltshire
Conservation awards, 154
  see also *Country Life* Farming and Wildlife Award;
    Silver Lapwing trophy
Conservation guidance, 39-41, 48, 67-70
Conservation in Agricultural Education Guidance Group, (CAEGG) 48, 52, 59, 123, 161, 163
Conservation Officer, 124
Conservation priorities, 176-8
Conservation teaching, 161-4
Control of Pollution Act, 96
Cornwall, 101
Cornwall FWAG, 100
Cornwallie, Dick, 10-11, 13
Corsham Court, 142
Council for the Protection of Rural England, (CPRE) 4, 12, 45, 86, 90
Council for the Protection of Rural Wales, (CPRW) 86, 120
Country Landowner, 40
Country Landowners' Association, (CLA) 2, 12, 14, 25, 33, 34, 37-8, 39-40, 42, 46, 47, 48, 56, 65, 88, 91, 111, 118, 120, 121, 132, 134, 151, 169
Country Life Farming and Wildlife Award, 128, 140, 180
  see also Silver Lapwing trophy

Index 197

Countryside, recreation in
    see recreation in the countryside
Countryside Act, 1968 10, 49
Countryside Adviser Scheme, 59,
    178-9, 181
    see also farm conservation advisers
Countryside advisers, 99
Countryside Campaigner, 183, 185
Countryside Commission, 4, 5, 8,
    10, 15, 25, 27, 31, 34, 35, 37, 39,
    45-6, 49-59, 60, 71, 88, 98, 112, 117-18,
    120, 122, 123, 134-5, 142-3, 147-8,
    149-50, 154, 167, 168, 174, 175-6,
    178-81, 186
    Monitoring Landscape Change
        Survey, 176
    Welsh Committee, 110, 111, 118-119
Countryside Commission for
    Scotland, 60
'Countryside in 1970' conferences, 3,
    8-11, 19, 40, 64
Countryside management, 49-50, 50-1,
    55, 180, 188
    see also management of existing
        features
Countryside Rangers, 179
Countryside Review Committee,
    39
County Agricultural Executive
    Committees, (CAECs) 64, 67
County Committees, 86, 87
County Councils, 86, 88, 94, 98
County Environment Groups, 64-7
County FWAGs
    see local groups
    see also individual entries
County Naturalists Trusts *see*
    Naturalists Trusts
County trusts for nature conservation,
    4, 10, 64, 88, 91, 98, 103, 132
Crane, Julian, 145-6, 147, 155-6, 158
Culling, Ted, 161, 162, 163
Cumbria FWAG, 81

Darke, Michael, 17, 20, 36-7, 64
Dartington Amenity Research Trust,
    (DART) 57, 77-9, 146
Dauntsey, Union Farm, 150, 164
Davidson, Joan, 176
Davidson, John, 52, 53
Dawson, Peter, 28, 29
Dawson, Wilf, 20, 27, 58, 68, 132, 185
Demonstration Farms, 50-1, 52, 71, 119,
    124, 130, 134-5, 148, 151, 153, 154, 163,
    168, 178, 179, 180
Department of the Environment,
    (DoE) 53, 189
Development Grants to Voluntary
    Bodies, 58
Devon FWAG, 81
Dinas Conference, 113-16
Dinas Exercise, 27
Divisional Surveyors, 182
Dorset FWAG, 81, 100, 187
Dutch Elm disease, 137, 142

East Suffolk Agricultural Institute,
    69-70
Ecological issues, 16-17
Economic and Social Research
    Council, 3
Economic Forestry Group, 117
Edmondson, Philip, 154-5
Education, 160-1
    see also Conservation in
        Agricultural Education
        Guidance Group;
        conservation teaching
Elliott, Fred, 48
Environmental movement, 8
Environmentally Sensitive Areas, 183,
    188, 189
Ernest Cook Trust, 76
Essex, 101
Essex Farmer, 80
Essex FWAG, 73, 79-81, 90, 123
Europe

see agricultural development in
      Western Europe;
      Common Agricultural Policy
European Conservation Year (1970)
  Conference, 68
Events
  see agricultural shows;
      meetings and events
Exercises
  see Chalklands Exercise;
      farming and wildlife exercises;
      Reaseheath Wetland Exercise;
      uplands exercise;
      wetlands exercise
Exmoor Society, 72

Fairbrother, Nan, 9
Farm conservation advice
  see Agricultural Development and
      Advisory Service;
      conservation advisers;
      farmers;
      Farming and Wildlife Advisory
      Group;
      Wiltshire
Farm conservation advisers, 179
  see also conservation advisers;
      farmers
Farm Diversification Scheme, 183
Farm Grants, 183
Farm survey work see survey work
Farm trails, 100, 151
Farm visits, 97, 103, 151-2, 153, 164, 169, 170
Farm walks, 26, 103, 113, 146
Farm walls, 189
Farm Woodland Scheme, 183, 188, 189
Farmers
  conservation advice for, 64-7, 93, 123, 149-54, 155, 164, 182
  decision-making by, 177
  liaison with, 67-70
  role of, in conservation, 188-9

  see also Farming and Wildlife
      Advisory Group (advisory
      capacity of);
      local groups;
      National Farmers; Union
Farmers' Union of Wales, (FUW) 90, 110-11, 117, 118, 120, 121, 123
  conservation attitudes of, 111-113
Farmers' Weekly, 14
Farming, Forestry and Wildlife
  Advisory Groups (FFWAGS) 5, 89
  study of, 110-24
  see also Wales;
      individual county FFWAGS
Farming with Wildlife, 11-12, 64, 97
*Farming and Wildlife: a Study in
  Compromise*, 17
Farming and Wildlife Adviser, 19-20
  see also National Adviser
Farming and Wildlife Advisory
  Group, (FWAG)
  advisory capacity of, 177-9, 180-1, 185-6
  aims of, 24
  attitudes towards, 42-6, 176-7
  charitable trust see Farming and
      Wildlife Trust
  county group network, 24
  development of, 2-3, 11, 14, 20, 24-60, 174, 178, 181, 186-90
  exercise, 25-32
  funding for, 32-5, 48, 51-3, 54-5, 56, 58, 60, 181, 186-8
  local contacts by, 64-7
  see also local groups
  Management Committee, 33-5, 74-5
  members of, 24-5
  National Committee, 72
  philosophy of, 139-141, 160, 163-4
  see also voluntary principle in
      conservation
  political context of, 35-48
  relationship of, with ADAS, 182-6

Index

relationship of, with CC, 49-59, 178-81
relationship of, with local authorities, 179-80
research project into, 3-5
role of, 9, 18-21, 47-9, 58, 174, 181, 182, 184-5, 188-9
Working Parties, 31, 70, 71
Farming and Wildlife Advisory Groups, (FWAGs) 2, 3, 4, 32, 47-8, 55, 58, 69, 189
county advisers, 5, 57, 139
county committees, 4-5
membership schemes, 103-4
see also individual county FWAGs; local groups
Farming and wildlife exercises, 25-32
Farming and Wildlife Trust, 5, 49, 58-60, 88, 93, 95, 96, 117, 123, 147, 148, 149, 178, 179, 185-6, 187, 189
Farms
  see Demonstration Farms; link farms
Ferguson-Lees, Ian, 20
FFWAGs *see* Farming, Forestry and Wildlife Advisory Groups
Field sizes, 13
Financial factors, 17
Fishbourne, Brigadier, 77
Fisher, James, 15
Fitton, Martin, 118
Food and Farming Year, 189
Food From Our Own Resources, 52
Food Production in the Countryside, 39
Footpaths, 66, 179
Forest Gate Farm, 165-6
Forestry, 113, 114, 115-16
  see also afforestation; tree planting on farms
Forestry Commission, 8, 11, 12, 31, 70, 98, 110, 113, 119, 120, 132, 143, 146
Forestry interests, 89
Forman, Dr. Bruce, 13

Friends of the Earth, 42, 90
Fuller, Ken, 134, 140-1, 143, 150
Funding *see* Farming and Wildlife Advisory Group
FWAG Guide to Priorities: Wildlife on the Farm, 177
FWAGs *see* Farming and Wildlife Advisory Groups

Game Conservancy, 90, 95, 146
  Cereal and Gamebirds Project, 169-70
Game conservation, 89-90, 141-2
Game Fair, 153
Gee, L., 20
Gillam, Beatrice, 132, 135-6, 144, 145, 146, 159
Glamorgan FFWAG, 122
Gloucestershire FWAG, 55, 59, 73, 76-7, 79, 131, 134, 138, 186-7
Gloucestershire Naturalists' Society, 76
Gloucestershire Trust for Nature Conservation, 76-7
Government attitudes, 41-2, 43, 46, 53-4, 55-6, 66-7, 174, 180, 188
Grant-aid schemes, 180, 183
Grants
  see Development Grants to Voluntary Bodies; Farm Grants; tree planting on farms
Grassland, 162
  see also chalk grassland; meadowland
Groundwork Foundation, 90
Gwent FFWAG, 117
Gwynedd FFWAG, 81, 100, 117, 122

Habitats, 16, 101, 103, 160, 177
Hall, Jim, 3-4, 20, 25, 27, 31, 32, 33, 34-5, 40, 42, 46, 48, 51, 57, 64, 65, 66, 67-70, 71-5, 75-6, 79, 81, 98, 120, 121, 130-3, 134-6, 145, 178

Hampshire FWAG, 83
Harper Adams Agricultural College, 137
Hedgelaying, 167
Hedgerow demonstration site, 164-5, 168
Hedgerow Management Demonstration Project, 166-7
Hedgerows and hedge planting, 13, 16, 65, 101, 102, 176, 189
Herefordshire FWAG, 82, 90, 100, 101
Hertfordshire, 14-15
Hertfordshire FWAG, 100
Hickling, R., 20
Hill-farming, 13, 113-14
Holdgate, Martin, 15
Honorary officers, (FWAGs) 91-3
Hookway, Reg, 51, 52, 53, 54, 56
Horton, Philip, 137
House of Lords
   debate, 43
   European Communities Committee, 128, 150
   Select Committee, 136, 139
Howatson, David, 120, 121-1
Hughes, Cledwyn, 64
Hughes, John, 55, 59, 76-7, 136, 152
Hughes, R.O., 113
Humberside FWAG, 69, 70, 72-4, 76

Ideology *see* politics and ideology in countryside issues
Insight, 38, 40
Institute of Terrestrial Ecology, 36, 95
Inter-agency relationships *see* Farming and Wildlife Advisory Group
Isaac, Bill, 134, 141-2, 145, 146, 158

Jealotts Hill Research Station, 95
Jeary, T.G., Ltd., 150
Jones, Sir Emrys, 15, 64, 113-14

Kegie, James, 118
Kent, 83

Kent FWAG, 59, 92, 100, 103
Kesteven Tree Society, 12
Kingston Deverill, 130, 134-5, 163, 168
Lackham College of Agriculture, 131, 132, 133, 134, 141, 146, 148, 151, 153, 154, 160-1
Laister, Geoffrey, 137, 138
Lancashire FWAG, 75, 90
Land and Water Service (LAWS) *see* Agricultural Development and Advisory Service
Land use, 27, 49-50
Landscape and agriculture, 9, 11
   see also Countryside Commission (Monitoring Landscape Change Survey)
Landscape conservation, 49, 53, 179, 181
Landscape Plan, 162
Lane, Stuart, 156
Laverack, Muriel, 54, 55
Lea, David, 13, 14, 15, 18, 19, 20, 27, 113
Leefe, John, 11-12
Leicestershire FWAG, 73, 100
Leonard, Patrick, 50, 147
Less Favoured Areas, 94
Liaison groups, 67, 69, 70, 74
Lincolnshire, 10, 12
   see also Lindsey Project for the Improvement of the Environment
Lincolnshire CPRE Trees Committee, 12
Lincolnshire FWAG, 100
Lincolnshire Trust for Nature Conservation, 10, 12
Lindsey and Holland Rural Community Council, 10
Lindsey County Council, 10, 12
Lindsey National Agricultural Advisory Service, (NAAS) 11
Lindsey Project for the Improvement of the Environment, (LPIE) 10-12

Index

Link farms, 52, 148, 154
Litter, 66
Livestock, 13
Lloyd, Richard, 142, 146, 147
Local authorities, 98, 179-80
Local groups
 advisory capacity of, 75-81
 county FWAGS:
  advice to farmers, 97-104
  advisers, 93-7
  composition and organization, 86-91
  effectiveness of advice, 104-5
  key honorary officers, 91-3
 network of, 81-3
 promotion of, 71-5
 survey of, 86, 105-6, 128-9
 see also individual county FWAGs
Lofthouse, Reginald, 20, 33, 35, 67
Lower Pertwood Farm, 135

McCarter, Nigel, 145, 146, 147
MacDonald, Richard, 60
Machin, William, 119
Management of existing features, 100-1, 102, 178
 see also countryside management
Management strategies, 14
Manpower Services Commission, (MSC), 123, 153, 166
 Special Measures Programme, 76
Marginal land, 16
Meadowland, 165-7
Meetings and events, 103-4
Melchett, Lord, 175-6
Methuen, Lord, 142
Ministry of Agriculture, Fisheries and Food, (MAFF) 2, 3, 4, 5, 12, 15, 18, 27, 33-4, 35, 38, 40-1, 45, 50, 52, 53, 54, 60, 67, 98, 110, 120, 131, 138, 152, 154-9, 167, 176, 183, 189
 advisory initiatives, 64-7
 see also Advisory Council for Agriculture and Horticulture; Agricultural Lane Service; Country Agricultural Executive Committees
Montgomery FFWAG, 89, 103, 117
 study of, 5, 119-124, 128
Montgomery Trust for Nature Conservation, 94, 122, 123
Montgomeryshire Countryside in 1970 Conference, 113
Moore, Dr. Norman, 13, 15, 20, 36-7, 46, 59, 67-8, 128, 176, 177
Morris, Peter, 160, 161

National Adviser, 81, 179, 182
 see also Farming and Wildlife Adviser
National advisory service, 49-60
 see also Countryside Adviser Scheme;
  farm conservation advisers;
  local groups
National Agricultural Advisory Service, (NAAS) 11, 12, 13, 15, 18, 19, 64, 66-7, 113
National Farmers' Union, (NFU) 2, 10, 12, 14, 15, 17, 18, 25, 33, 34, 37, 38, 39-40, 42, 43, 47, 48, 56, 59, 60, 65, 67, 70, 72, 73, 88, 90, 91, 98, 118, 120, 121, 123, 131, 132, 134, 167, 175, 176, 179
 Welsh Council, 110-11
National park authorities, 90
National Parks, 11, 124, 182
National Trust, 15
National Union of Agricultural and Allied Workers, 65
Naturalists Trusts, 68, 69, 72, 123
'Nature Conservancy and Agriculture, The', 67-8
Nature Conservancy Council, (NCC) 2, 4, 8, 13, 15, 25, 28, 34, 35-7, 39, 50, 53, 54, 56, 58, 60, 64, 69, 70, 72, 76,

88, 116-17, 118, 119, 120-1, 122, 132, 134, 157, 159, 167, 179, 180, 181
  Toxic Chemicals and Wildlife Section, 36, 68
  Welsh Committee, 110
Nature conservation
  see conservation;
    county trusts for nature conservation
Nature Conservation and Agriculture, 37, 39
Nelson, Dorothea, 76, 77-8, 136
New Agricultural Landscapes, 37, 39, 50, 51, 180
New features, creation of, 100-1
New Lives, New Landscapes, 9
New Scientist, 16
Newbery, Peter, 131, 133
Newell, Peter, 131, 132
Newsam, Stuart, 133
Newsletter, 95-6
Norfolk FWAG, 75, 83, 177
North Humberside see Humberside FWAG
North Wales Naturalists' Trust, 120
North York Moors, 100, 105
Northumberland Countryside Liaison Group, 74
Northumberland FFWAG, 74, 89, 92, 100

Oliver-Bellasis, Hugh, 169
Open days, 103
Organic farming, 169
Organizations represented on FWAGs, 86-91
Ornithological societies, 89
  see also British Trust for Ornithology; Royal Society for the Protection of Birds
Osborn, Alison, 59, 139, 145, 149, 150-4, 160-1, 164-70
Overend, Eunice, 157, 158, 159

Oxford Farming Conference, 189
Oxfordshire FWAG, 80, 100

Pembrokeshire FFWAG, 117, 122
Pesticides, 9, 13, 16
Phillips, Adrian, 46
Phillipson, Peter, 140
Phoenix, 146
Planning and the Countryside, 47
Planning controls, 42, 160
Planning department, county, 4, 8
Plumb, Henry, 15, 68
Politics and ideology in countryside issues, 81, 174, 175
  see also voluntary principle in conservation
Pond creation, 100, 101, 102, 167, 177
Potter, Clive, 140, 177
Powys County Council, 120
Prestt, Ian, 51-2, 74
Prince of Wales Committee, 113
Priorities see conservation priorities
Pritchard, Tom, 115, 119, 121
Pwllpeiran, 115

Qualifications and experience of advisers, 94-5
Quangos, 41, 56

Radnorshire, 83, 117, 122, 123
Ramblers' Association, 42, 90
Ratcliffe, Bill, 117
Rayner, Sir Derek, 41
Reading University College of Estate Management, 33
Reaseheath Wetland Exercise, 30
Recreation in the countryside, 19
Rees, *Meuric*, 116
Rice, David, 133, 136, 137, 138, 142-5, 147, 150, 151, 152, 153, 154, 159, 161, 162, 167
Riddick, Peter, 131
Roberts, Meredydd, 115
Royal Agricultural College, Cirencester, 149

Index

Royal Forestry Society, 95
Royal Institution of Chartered Surveyors, 31, 33, 86
Royal Society for Nature Conservation (RSNC), 2, 10, 42, 91
Royal Society for the Protection of Birds, (RSPB) 2, 3, 4, 13, 14, 18, 19, 20, 32, 33, 34, 44-5, 51-2, 72, 88-9, 116, 120
'Rural advisers', 37
Rushall, 169

Scotland, 82
Secretary of State for the Environment, 55, 56
Secretary of State for Wales, 110
Selly, Clifford, 14
Set-Aside, 183, 188
Shelbourne, Earl of, 164-5
Shoard, Marion, 42
Shooting Times, 16
Shropshire FWAG, 177
Silsoe Committee, 17-20
Silsoe Exercise, 3, 8-21, 113, 114, 116
   background to, 12-14
   conference, 14-17, 24
   post-conference, 17-21, 25-6, 64
Silver Lapwing trophy, 128, 138
   see also *Country Life* Farming and Wildlife Award
Sites of Special Scientific Interest, (SSSIs) 37, 118, 129, 179, 182
   see also Berwyn SSSI
Skelmersdale, Lord, 43
Small, Colin, 121, 122
Smith, A.E., 10
Smith, Walter, 15
Society for the Promotion of Nature Reserves (SPNR), 2, 10, 18, 33, 51, 64, 132
Soden, Ronald, 113, 117, 119
Soil deterioration, 9
Somerset FWAG, 76, 79, 100
Somerset Levels, 159

Somerset Trust for Nature Conservation, 76
Spink, Neville, 132-3, 134, 135, 136, 138, 154, 158, 159
Sports Council, 184
Staffordshire FWAG, 83, 93
Stratton, David, 130, 134-5, 140, 150, 155, 164
Stratton, Michael, 27, 130, 132, 133
Structure plans, 8
Strutt, Sir Nigel, 41
Strutt Report, 53-4, 112
Suffolk Countryside Committee of FWAG, 70
Suffolk FWAG, 69-70, 77, 78, 79, 177
Suffolk Naturalists' Society, 70
Suffolk Trust for Nature Conservation, 70, 77
Suggett, Graham, 48, 161
Surrey FWAG, 73
Survey work, 100, 102, 166
Surveying expertise, 155
Sussex FWAG, 103, 121
Syndicate groups, 114-15

'Take a Pride in Your Landscape', 13
Technological change and responses to, 9
Theft of the Countryside, The, 42
Tickler, Gordon, 20
Timber Growers Organisation, 120
Times, The, 29, 142
Tinker, Jon, 16
Transport and General Workers Union, 65
Tree planting on farms, 11-12, 13, 16, 51, 66, 98-100, 101, 102, 103, 141-2, 176-7
   grants for, 142-4
   see also Farm Woodland Scheme
Tree Planting Scheme *see* Wiltshire
'Trees on Farmland' conference, 137
'Trees on the Farm' seminar, 12
Trist, John, 13, 20, 69
Turner, Keith, 34, 57, 135, 142, 143

Tyne Tees, 100
Ulster, 82
Union Farm, Dauntsey *see* Dauntsey, Union Farm
Uplands exercise, 27
Urban growth, 15

Voluntary principle in conservation, 2, 39-40, 46, 47, 55, 56-7, 59, 129, 159, 160, 174-8, 188

W.A. Cadbury Trust, 34
Wales, 5, 27, 60, 82, 83, 86, 88, 90, 93
   study of FFWAGs in, 110-24:
      Dinas Conference 113-16
      Dinas to FFWAG 116-18
      FUW and conservation 111-13
      Montgomery FFWAG 119-24
      Wales Countryside Forum 112, 118-19, 121
   see also Farmers' Union of Wales; Farming, Forestry and Wildlife Advisory Groups; individual counties
Wales, HRH The Prince of, 59
   see also Prince of Wales Committee
Wales Countryside Forum, 112, 118-19, 121
Walker, Peter, 35
Wallace, David, 20
Walters, Peter, 132, 133, 160
Warwickshire College of Agriculture, 48
Water authorities, 66, 86, 89
Watkins, Tom, 121
Webb, Ronald, 119, 120, 121
Weed Research Organization, 95
Welsh Office, 53, 110
Welsh Office Agriculture Department, (WOAD) 110, 116-17
Welsh Woodlands Group, 119
Wetlands exercise, 27, 30, 31
   see also Reaseheath Wetland Exercise

Whitlock, Ralph, 29
Wilder, Bill, 128, 133, 134, 136-40, 145, 146-7, 148, 150-1, 155, 158, 160, 169
Wilder, Jean, 137, 140
Wildlife, 13, 14, 16, 38-9, 114-16, 146, 179, 181
   see also farming and wildlife exercises
Wildlife and Countryside Act, 2, 3, 43, 47, 60, 81-3, 112, 159, 160, 174-5, 180, 182, 188
Wildlife and Countryside Bill, 42, 45, 55-6, 58, 160
Wildlife Link
   Badger Working Group, 158
Wildlife Survey Team, 28
Williams, Stephen, 121
Williamson, Kenneth, 13, 20, 29-30
Wiltshire, 5, 27, 28
   Tree Planting Schemes, 142-3
   see also Community Council for Wiltshire
Wiltshire County Council, 146-7, 149, 167-8
Wiltshire County Planning Department, 134
   Forestry Adviser, 133
   Officer, 131
Wiltshire FWAG, 59, 98
   study of FWAG in, 86, 128-70
      activities, 150
      advisory role, 165, 166, 167
      advisory team and Adviser, 144-9, 149-54, 160-1, 164-70
      budget, 168
      Chairman and philosophy of, 136-42
      Chalklands Exercise, 133-6
      committee members, 133-4
      establishment of, 129-33
      fund raising, 149-50, 168
      office costs, 154-4
      policies and attitudes, 155-7

review of progress, 164-70
Steering Group, 150, 152, 153, 154, 157-8
trainee, 153, 154
trees and grants, 142-4
Wiltshire Natural History Forum, 131-3, 134, 145
Wiltshire Ponds and Lakes Association, 167
Wiltshire Radio, 149
Wiltshire Trust for Nature Conservation, 29, 129-33, 134, 138, 140, 145, 146, 149, 167
Field Officer, 165-6, 176-7
Winnifrith, Sir John, 15
Women advisers, 96-7
Womens' Institutes, 77
Woodland management, 137-8, 141-2
see also Farm Woodland Scheme; tree planting on farms

Woodland Trust, 51, 57
Woodlands
  see Coed Cymrus;
    forestry;
    Forestry Commission;
    Welsh Woodlands Group;
Woodman, Michael, 68, 69
Wookey, Barry, 169, 170
Worcestershire FWAG, 103-4
World Wildlife Fund, 34

Y Tir, 113
York-King, Tom, 169
Yorkshire, 69
  see also North York Moors
Yorkshire FWAG, 70, 73, 100
Young Farmers' Clubs, 77, 86, 123, 134, 154
  National Federation of, 149